Reduction of a Ship's Magnetic Field Signatures

Reduction of a Ship's Magnetic Field Signatures
John J. Holmes

ISBN: 978-3-031-00580-0   paperback

ISBN: 978-3-031-01708-7   ebook

DOI: 10.1007/978-3-031-01708-7

A Publication in the Springer series

SYNTHESIS LECTURES ON COMPUTATIONAL ELECTROMAGNETICS # 23

Series Editor: Constantine A. Balanis, Arizona State University

Series ISSN
ISSN 1932-1252   print
ISSN 1932-1716   electronic

Reduction of a Ship's Magnetic Field Signatures

John J. Holmes
Naval Surface Warfare Center

SYNTHESIS LECTURES ON COMPUTATIONAL ELECTROMAGNETICS # 23

iv

ABSTRACT

Decreasing the magnetic field signature of a naval vessel will reduce its susceptibility to detonating naval influence mines and the probability of a submarine being detected by underwater barriers and maritime patrol aircraft. Both passive and active techniques for reducing the magnetic signatures produced by a vessel's ferromagnetism, roll-induced eddy currents, corrosion-related sources, and stray fields are presented. Mathematical models of simple hull shapes are used to predict the levels of signature reduction that might be achieved through the use of alternate construction materials. Also, the process of demagnetizing a steel-hulled ship is presented, along with the operation of shaft-grounding systems, paints, and alternate configurations for power distribution cables. In addition, active signature reduction technologies are described, such as degaussing and deamping, which attempt to cancel the fields surrounding a surface ship or submarine rather than eliminate its source.

KEYWORDS

Underwater electromagnetic ship signatures, ship signature reduction, signature compensation, magnetic silencing, degaussing, deamping

Acknowledgment

The author sincerely appreciaes the many helpful comments of Mr. J. Scarzello during the review of this manuscript. Also, the people who have dedicated their careers developing the technologies described in this manuscript deserve all our thanks.

Dedication

To Mom and Dad

Contents

1. Introduction 1

2. **Passive Magnetic Silencing Techniques** 7
    2.1 Passive Reduction of Ferromagnetic Signatures 7
    2.2 Passive Reduction of Roll-induced Eddy Current Signatures 21
    2.3 Passive Reduction of Corrosion-related Magnetic Field Signatures 30
    2.4 Passive Reduction of Stray Field Signatures 36

3. **Active Signature Compensation** 43
    3.1 Degaussing System Design 43
    3.2 Degaussing Coil Calibration and Control 49
    3.3 Active Reduction of Corrosion-related Magnetic Field Signatures 56
    3.4 Closed-loop Degaussing 58

4. **Summary** 63

CHAPTER 1

Introduction

Reducing the magnetic field signature of a naval vessel will also decrease its susceptibility to actuating naval influence mines and, in the case of a submarine, its probability of detection by underwater barriers and maritime patrol aircraft. The application of naval technologies to achieve these goals is sometimes called *magnetic silencing*. Decreasing a vessel's magnetic signature below the detection threshold of either an influence mine or a submarine detection system's inherent noise level will negate them as threats. However, important military benefits can still be realized even for signature reduction levels less than 100%.

Decreasing a naval vessel's magnetic signature has a synergistic payoff against the threat of naval influence mines when used in combination with hunting and sweeping. If ship signatures are lowered to a level where influence bottom mines cannot detect them in deeper waters, then those military-contested areas need not be hunted or swept for these mine types before combatants enter. This could save a significant amount of time during forced-entry operations or could substantially reduce dedicated mine countermeasure resources needed to successfully complete the mission. After the conflict, all areas will eventually have to be cleared of mines to allow the safe passage of commercial shipping not equipped with magnetic silencing systems.

In shallower waters, reducing magnetic signatures lessens the effective density of the minefield [1]. If ship signature amplitudes are decreased, the planner of a shallow-water minefield must increase the sensitivity of the deployed mines so as not to miss target vessels that pass within the damage radius of the weapon's explosive charge. Failure to increase the mines' actuation sensitivity could result in a catastrophic failure of the entire minefield, resulting in little or no threat to transiting vessels. Conversely, if the minefield planner does increase the mines' sensitivity to counter the reduction in target vessel signature levels, then minesweeping systems become more effective against the sensitive weapons. In this case, swept widths increase against the more sensitive mines, requiring fewer passes of the minesweepers (less time) to achieve a desired level of risk to the follow-on task force, or fewer sweeping platforms are necessary to successfully complete the mission. Finally, naval vessels equipped with advanced signature compensation systems have the ability to alter the spatial and temporal characteristics of their magnetic fields and generate a signal that might jam a naval mine. Mine jamming is a countermeasure tool that prevents an influence mine from coming to a firing decision point in its preprogrammed logic as the protected vessel sails safely over the weapon.

Due to the stealthy mission requirements of submarines, minesweeping is generally not an option to counter an antisubmarine minefield. The loss of sweeping as a mine countermeasure tool requires submarine signature levels to be much lower than surface ships under similar scenarios [1]. In this case, mine jamming may also mitigate some of the increase in submarine mine susceptibility caused by the elimination of sweeping, enabling them to penetrate minefields while maintaining their stealthy advantage.

Poor acoustic conditions in shallow littoral ocean environments have increased the importance of detecting submarines by their electromagnetic field signatures. In the past, large inductive loops have been installed on the seafloor as underwater magnetic barriers to prevent submarines from covertly entering protected ports, harbors, and other important naval installations [1]. With modern field sensing technology, portable underwater magnetic barriers for detecting submarines can be deployed in forward areas. In addition, miniaturization of low-power high-sensitivity magnetic field sensors is enabling the development of magnetic anomaly detection (MAD) systems that can be installed on inexpensive long-duration unmanned air vehicles (UAVs) [2]. Deploying "swarms" of these MAD equipped UAVs, which are controlled in a cooperative behavior search pattern, could monitor large shallow-water areas of the ocean and detect acoustically quiet submarines.

The shift in naval operations away from deep water and more toward shallow (littoral) ocean environments has increased the importance of controlling a surface ship and submarine's magnetic field signature. The four primary ship-related sources of magnetic field in the ultralow frequency (ULF) band [approximately zero to 3 Hz] are:

1. Ferromagnetism induced by the Earth's natural magnetic field in the ferrous steel used to construct naval vessels;
2. Eddy currents induced in any shipboard electrically conducting material (magnetic as well as nonmagnetic) as it rotates in the Earth's magnetic field;
3. Electric currents impressed into a ship's conducting hull and the surrounding seawater by natural electrochemical corrosion processes or by cathodic protection systems designed to prevent the ship from corroding (rusting);
4. Currents that flow in electric motors, generators, distribution cables, switch gear, breakers, and other active circuits found onboard.

Although an in-depth description of these sources of magnetic field is given by Holmes [1], a brief summary of the physical processes behind each is presented here in order understand the magnetic signature reduction technologies to be discussed later.

The most important shipboard source of magnetic field is the magnetization of ferromagnetic steel used in the construction of a naval vessel's hull, internal structure, machinery, and equipment items. The ferromagnetic source can be further divided into two components called the induced and permanent magnetization. The ship's induced signature is the result of the Earth's naturally uniform magnetic field being distorted by the steel used in the vessel's construction. This anomaly in the Earth's field is detected as a ULF signal by a mine or barrier's stationary magnetic field sensor as the vessel sails past it or by an airborne MAD sensor as it flies across a submerged submarine. The vessel's induced magnetization and associated off-board signature will change with its roll-and-pitch angle, heading, latitude, and longitude. As the ship is mechanically stressed, part of the induced magnetization is retained as permanent or residual magnetization that generally changes slowly over time. The ferromagnetic components typically dominate a surface ship or submarine's signature and must be addressed first in the design of a magnetic silencing system.

Eddy currents are induced in electrically conducting materials onboard surface ships principally as they roll within the Earth's magnetic field. Ships constructed from conducting metals, such as aluminum, stainless steels, or titanium, will generate eddy current magnetic signatures although these materials are themselves nonmagnetic. The fields produced by eddy currents can be large enough to detonate mines and are the second most important shipboard source of magnetic signature.

The third largest, and least known, of the major sources of magnetic field is corrosion currents that flow in and around a surface ship or submarine's hull. When a vessel's steel hull is electrically connected to its nickel–aluminum–bronze propeller and immersed in seawater, a battery is formed. The primary path for corrosion currents is from the ship's hull through the seawater to its propeller or propellers, then up the shaft through the bearings and drive mechanism, and eventually back to the hull to complete the electric circuit. Corrosion currents are a source of both static and alternating magnetic field signatures.

Cathodic protection systems are used to prevent a ship's metallic hull from corroding. The principle of their operation is to turn anodic materials into cathodes. The two types of cathodic protection systems in use on modern naval vessels are the passive cathodic protection system and the impressed current cathodic protection (ICCP) system. A passive cathodic protection system is composed of a large number of zinc bars that are welded to the hull. The electrochemical potential of zinc is more negative than steel and serves as an anode when attached to the hull. With the zinc in place, the hull is turned into a cathode and protected from rusting. The zinc bars themselves will corrode and must be replaced periodically, giving this passive method of cathodic protection the name sacrificial anode system.

An ICCP system is used primarily on large ships. Instead of zinc bars, ICCP system anodes are made of platinum-coated wires or rods that are mounted on the hull inside an insulated housing. The anodes are wired to internal power supplies whose return leads are grounded to the hull. The

ICCP anodes actively pump current into the seawater, turning the hull once again cathodic. The voltage at the ICCP anodes must be constantly regulated to ensure that sufficient current is flowing to protect the ship from corroding while not allowing too much current to enter the hull that might cause hydrogen embrittlement and weaken it. Silver–silver chloride electrodes called reference cells are mounted at several positions on the hull to monitor the effects of the anode current and to regulate it accordingly. A ship's ICCP system automatically adjusts its anode current until the reference cells measure a specified potential relative to the hull, called the set potential. Generally, the set potentials for naval ICCP systems range from about –800 to –850 mV with respect to the hull.

Cathodic protection systems, especially ICCP types, can drive large amounts of current into the sea which then flow through the hull, mainly parallel to its longitudinal axis. The hull and propeller shaft current are the main sources of off-board corrosion-related magnetic (CRM) signatures. As the shaft rotates, the variable contact resistance between it and the bearings modulates the corrosion currents. The shaft-modulated currents excite alternating CRM fields that occur at the fundamental shaft rotation frequency, plus harmonics. The ship's CRM source can be represented as a DC and alternate current (AC) longitudinal electric dipole with its magnetic fields circulating around it according to the right-hand rule.

The last of the major shipboard sources of magnetic field are called stray field sources. Stray field signatures are produced by any current carrying electric circuit found onboard a ship. The larger of the stray fields are produced by the vessel's electromechanical machinery and power distribution system. High-power electric generators, motors, switchgear, breakers, and the distribution cables that interconnect them can emit both DC and AC fields.

The magnitude and importance of magnetic stray field signatures will increase in the near future. Trends toward constructing naval vessels from nonmagnetic metals, such as aluminum or stainless steel, could reduce the shielding effectiveness of the hull. Also, the U.S. Navy has committed to developing an "all electric" ship that will use large electric motors for propulsion. Since the power supplied to electric propulsion motors could exceed 30 MW, very high voltages and, more importantly, very large currents would be flowing inside the ship's power system. The problem would be exacerbated if the motors are mounted exterior to the ferrous hull where no shielding at all would be present. Both DC and AC stray field signature components must be combined with the other three sources in assessing a vessel's true susceptibility to magnetic field detection.

Almost every aspect of ship and ship system design can affect underwater electromagnetic field signatures. Hull shapes and their roll characteristics, along with bulkhead and deck geometries, will impact the ferromagnetic and eddy current field components of a vessel. The magnetic and electric properties of all materials selected for use in the construction of ships and submarines are a major consideration with respect to ferromagnetic, eddy current, and CRM signatures. Propeller and propulsion system designs, combined with the vessel's cathodic protection system arrangement, impact the resultant corrosion-related signatures. Approaches taken in the design of high-power (high-current) electric motors, generators, and distribution subsystems and components can exacerbate or conversely eliminate stray field signatures. The development of underwater electromagnetic signature reduction technologies is not an isolated process separate from the design of the overall ship and its systems, but must be considered synergistically to realize maximum silencing performance at minimum cost.

The underlying physics of shipboard sources has to be understood to design signature reduction technologies that are low cost with a minimum impact on the vessel, its systems, and their operation. The first rule in the development of any signature reduction system is to eliminate as many of the sources as is technically feasible and affordable before attempting to actively cancel the remaining fields. For example, constructing naval vessels from nonmagnetic and nonconducting materials could yield a 40-dB reduction in its total magnetic field signature produced by ferromagnetic, eddy current, and CRM sources. (Some ferrous steel that is part of machinery items, weapons, and other ship systems may have to be retained for them to operate properly.) In addition, care in the up-front design of high-power electric propulsion motors, generators, and their interconnecting distribution system could trim down a large portion of the stray magnetic fields cheaply and with little impact on the ship. Any magnetic signatures that remain after the source elimination process could be reduced by another 20–40 dB through active field cancellation techniques.

An active magnetic compensation system attempts to artificially generate a signature that is identical in amplitude and shape to the vessel's uncompensated field, but of opposite polarity. The superposition of the ship's compensating field with its uncompensated flux pattern results in a small net magnetic signature. Active compensation of a naval vessel's ferromagnetic, eddy current, and stray fields is achieved with a controlled array of onboard magnetic sources called a degaussing system. The CRM signature is actively cancelled with controlled electric current sources placed along the ship's hull and is called a deamping system. One drawback of active field cancellation is its requirement for a monitoring system that can detect changes in the vector components of onboard sources so that the compensation system can be readjusted to maintain a low signature.

This monograph will introduce the reader to both passive and active magnetic silencing techniques and simple computational models used to study their effectiveness. The benefits of changing the magnetic properties of hull materials will be demonstrated with a prolate spheroidal shell model of induced magnetization. A cylindrical shell mathematical model will be derived and exercised to demonstrate eddy current signature reduction levels that might be achieved through the use of less electrically conductive hull materials and with ship roll stabilization. Techniques to reduce the CRM and stray fields can be sufficiently demonstrated with simple dipole models of these sources. Stability issues associated with the adjustment (calibration) of an active signature compensation system will be examined mathematically along with techniques to regularize the process. Finally, underwater sensor ranges and fixed facilities used to periodically monitor and calibrate compensation systems will be covered briefly, along with an onboard self-monitor system called closed-loop degaussing.

REFERENCES

[1] J. J. Holmes, *Exploitation of a Ship's Magnetic Field Signatures*, 1st edn. Morgan & Claypool Publishers, Denver, CO, 2006. doi:10.2200/S00034ED1V01Y200605CEM009

[2] R. Tiron (2006, April). Gulf Nation Poised to Lead Region in Production of Unmanned Aircraft. National Defense Industrial Association. Arlington, VA. [Online]. Available: http://www.nationaldefensemagazine.org/issues/2005/Apr/Gulf_Nation.htm.

• • • •

CHAPTER 2

Passive Magnetic Silencing Techniques

2.1 PASSIVE REDUCTION OF FERROMAGNETIC SIGNATURES

The two attributes of a naval vessel that affect its ferromagnetic signature are its size and the permeability of the material used in its construction. Traditionally, ships and submarines have been designed with length-to-beam ratios on the order of 10:1 to accommodate hydrodynamic requirements. Today, nonconventional hull geometries with length-to-beam ratios of 4:1 are being constructed [1]. However, this change in length-to-beam has little impact on the magnetic signature assuming the vessel's volume and magnetic material properties are held constant. This can be demonstrated with a simple example.

As explained by Holmes [2], a ship or submarine's far-field magnetic signature is directly proportional to its dipole moment. Modeling the vessel as a right circular cylindrical with a magnetic permeability much greater than free space ($\mu_0 = 4\pi \times 10^{-7}$ Henrys/m), the equivalent longitudinal magnetic dipole moment, m_1, and transverse moment, m_t, can be approximated as [3]:

$$m_1 = \alpha_1 H_1 \qquad m_t = \alpha_t H_t \tag{2.1}$$

where α_1 and α_t are the longitudinal and transverse magnetic polarizabilities of the cylinder, and H_1 and H_t are the Earth's inducing field in the cylinder's longitudinal and transverse directions, respectively. For a length-to-beam ratio of 10:1, α_1 and α_t are given by Fogiel (2007) as −1.06 and −1.94 times the cylinder's volume, whereas for a 4:1 ratio, α_1 and α_t are −1.16 and −1.85 times the volume, respectively. Under the assumption that the vessel's volume is the same at both length-to-beam ratios, the change in magnetic dipole moment and the resultant far-field signature are negligible.

As implied from the example above, a ship or submarine's far-field magnetic signature will proportionally decrease with its size or volume. In the past, this approach was generally not an acceptable option for reducing magnetic field signatures, since less ship size translates into less payload capacity. However, the present trend of building faster surface combatants with smaller hulls has the added benefit of a lower magnetic signature. For example, the uncompensated magnetic signature of a littoral combat ship (LCS)-size steel hull combatant is expected to be almost three times smaller than a DDG 51 class destroyer.

One geometrical aspect of a ship's magnetic hull that does significantly impact its off-board signature is its thickness. The equations for the induced longitudinal magnetic (ILM) field of a prolate spheroidal shell were given by Holmes [2] and will be used here to demonstrate the relationships between a hull's thickness, magnetic permeability, and its induced signature. Due to the nature of the prolate spheroidal coordinate system, the shell's thickness will be smaller on its ends compared to its middle. Therefore, hull thicknesses will be specified at both locations in this example.

Over the range of values typical for naval vessels, off-board magnetic signatures are nearly proportional to the hull's thickness. Plotted in Figures 2.1 and 2.2 are the vertical and longitudinal fields for a prolate spheroidal shell with thicknesses on its ends/middle (bow/beam) of 0.5/3, 1/5, 1.5/8, and 2/10 cm. In all cases the outside dimensions of the hull are fixed at 100 m in length and

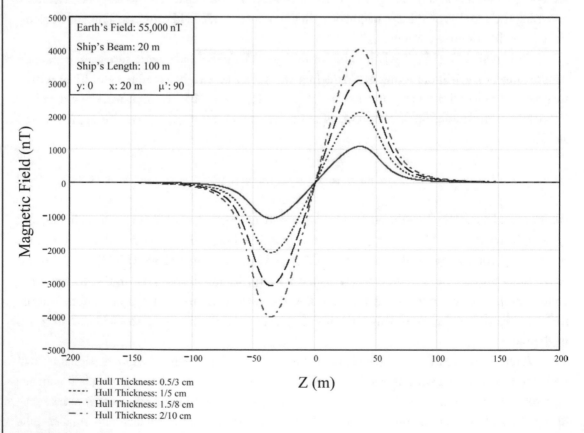

FIGURE 2.1: Vertical magnetic field signatures produced by the induced longitudinal magnetization of prolate spheroidal shells of varying thicknesses.

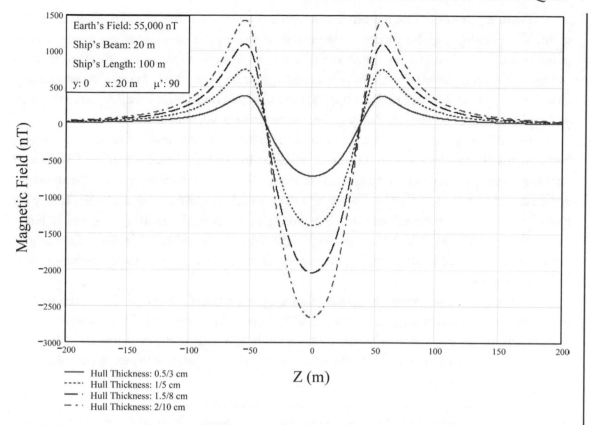

FIGURE 2.2: Longitudinal magnetic field signatures produced by the induced longitudinal magnetization of prolate spheroidal shells of varying thicknesses.

20 m in diameter, and with a magnetic permeability constant ($\mu' = \mu/\mu_0$) set equal to 90 (HY80 steel). The inducing field along the spheroid's longitudinal axis was selected as 55,000 nT, whereas the signatures were computed 20 m directly below the spheroid's axis. As the figures show, a factor of 3 to 4 reduction in hull thickness will also reduce the magnetic signature amplitudes by a similar amount.

The design parameter that has the greatest impact on passively reducing a ship or submarine's magnetic field signature is the permeability constant of the material used in its construction. As discussed by Holmes [4], naval vessels have historically been made out of armor steel plating to protect them from artillery fire, and in modern times, the armor reduces their vulnerability to missiles, torpedoes, and sea mines. The primary element alloyed in steel armor plate is iron, the source of its ferromagnetism. However, modern metallurgical technologies have produced suitable nonmagnetic alternate materials that are being used in the construction of naval combatants.

Ferromagnetism in iron is produced by spinning electrons in the 3d orbital shell of the element. The spinning negatively charged electrons produce a dipolar magnetic source whose axis can point in either of two directions, called up or down spin. In atoms of ferromagnetic elements, such as iron, cobalt, and nickel, all orbital shells are filled with equal numbers of up and down spinning electrons except for the 3d. The four unpaired electrons in iron's 3d orbit have a net nonzero magnetic spin moment and can influence unpaired 3d electrons in adjacent atoms.

In addition to having unpaired electrons in its 3d orbit, neighboring atoms of ferromagnetic materials must be spaced within its crystalline structure at distances that are favorable to exchanging energy between their unpaired electrons and affect each other's spin. If the atoms are spaced too close together, they have a negative energy exchange and are nonmagnetic. If they are spaced too far apart, their influence on neighboring atoms is small, resulting in a weakly ferromagnetic material. Only those elements with unpaired electrons in the 3d orbits of their atoms, which are also spaced in the crystal at the proper distances for a positive energy exchange, tend to be ferromagnetic.

Alloying of elements can change their ferromagnetic properties through adjustments in their crystalline structure. If manganese is alloyed with copper, aluminum, and tin, its atomic spacing is increased, producing a ferromagnetic compound although none of the constituent elements are by themselves ferromagnetic. Conversely, if iron is alloyed with higher amounts of chromium and

TABLE 2.1: Magnetic permeability of ship construction materials	
MATERIAL	μ
High-strength steel	180
HY80 steel	90
Cold-rolled 304 stainless steel	10
AL6XN stainless steel	1.01
EN 1.3964 stainless steel	1.01
Pure aluminum	1.00
Pure titanium	1.00
Wood	1.00
Carbon fiber	1.00

nickel, the resulting steel can be made nonmagnetic since its atomic configuration does not support a favorable exchange of energy between atoms.

Steels forged with larger amounts of chromium are called stainless steels, are very resistant to corrosion, and can have a very low permeability constant. Many types of stainless steels are produced with varying properties depending on the proportions of iron, chromium, nickel, carbon, nitrogen, and other elements used in the alloying process. However, not all stainless steels are nonmagnetic. Some martensitic stainless (higher carbon stainless steels) can still be ferromagnetic, whereas some austenitic steels (higher chromium content) can have a very low magnetic permeability. When cold-worked or cold-welded, some austenitic steels, such as 304 stainless, form martensitic pockets that increase their overall magnetic permeability.

Naval vessels have and are being built out of materials that have wide variations in their magnetic properties. Table 2.1 lists several of them along with their relative permeability. Although

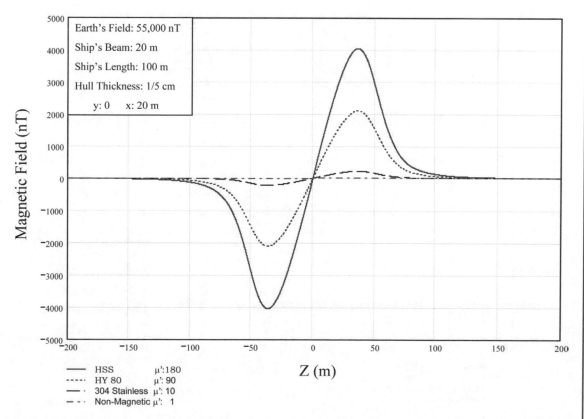

FIGURE 2.3: Vertical magnetic field signatures produced by the induced longitudinal magnetization of prolate spheroidal shells of varying permeability.

the permeability constant given for high-strength steel (HSS) is typical of modern armor hull plate, values approaching 300 have been measured, while HY80 steel seems to have less variation about the stated value [5]. Conversely, 304 stainless steel has a permeability constant close to 1 when it is initially forged and can approach 10 after cold rolling [6]. Two of the new super austenitic stainless steels are listed in Table 2.1, namely, AL6XN and its European equivalent EN 1.3964. Both steels are very resistant to corrosion and have a relative permeability that is effectively 1. (A technical data sheet for AL6XN can be found in Fogiel [7], while the properties of EN 1.3964 are given by Fogiel [8].) The last six materials listed in Table 2.1 may all be considered nonmagnetic.

Constructing ships from materials with low magnetic permeability can have significant pay-offs in passively reducing their off-board signatures. This can be demonstrated with the prolate spheroidal shell example from above. Using bow and beam hull thickness of 1 and 5 cm, respectively, the vertical and longitudinal magnetic field signatures are plotted in Figures 2.3 and 2.4 for HSS,

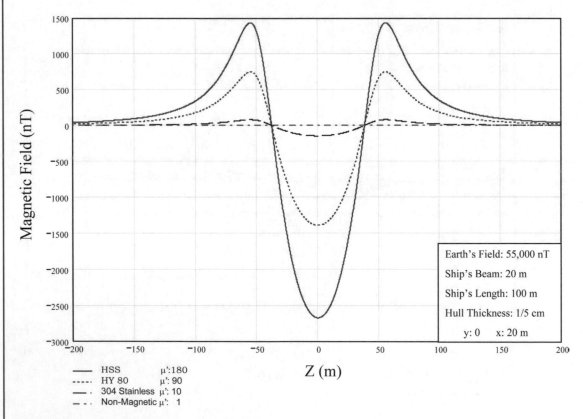

FIGURE 2.4: Longitudinal magnetic field signatures produced by the induced longitudinal magnetization of prolate spheroidal shells of varying permeability.

HY80 steel, cold-worked 304 stainless steel, and the trivial case of nonmagnetic materials ($\mu' = 1$). As these data show, the best technique for significantly reducing the static magnetic field signatures of naval vessels is through the use of nonmagnetic materials in their construction.

Several classes of naval ships and submarines have or are presently being built from nonmagnetic materials. In the United States, the high-speed vessel (HSV) surface ship and one LCS design [1] shown in Figure 2.5 are constructed mainly from aluminum, while the German 212 submarine is made from EN 1.3964 [8] and the Russian Alpha class submarine is made from titanium [9]. The hull of the U.S. MCM-1 class minesweeper is wood and fiberglass (Figure 2.6), whereas the Swedish Corvette HMS *Visby* has a carbon fiber-based construction [10]. The composite hulls of the latter two examples are not only nonmagnetic but also have very high electrical resistances in comparison to metallic hulled vessels, which minimizes their roll-induced eddy current-generated magnetic signatures to be discussed later.

a.) United States HSV

b.) United States LCS

c.) German 212

d.) Russian Alpha

FIGURE 2.5: Examples of naval vessels constructed from nonmagnetic materials.

a.) *USS AVENGER* (MCM-1)

Wood & Fiberglass Hull

Length: 68 meters

Displacement: 1300 tons

Speed: 14 knots

b.) *HMS VISBY*
(Swedish Corvette)

Carbon Fiber Hull

Length: 73 meters

Displacement: 600 tons

Speed: 35 knots

FIGURE 2.6: Examples of naval vessels constructed from nonconducting materials.

Although ship hulls, bulkheads, and decking can be made from nonmagnetic materials, some internal shipboard items must still be constructed from ferromagnetic steels for them to operate properly. Machinery, such as engines and portions of the ship's propulsion train, must be made from ferromagnetic steels to operate reliably within a high-temperature and high-stress environment. This also includes components found within guns and other weapons systems. In addition, electro-mechanical motors and generators along with power distribution equipment, such as transformers and circuit breakers, need to be constructed in part from ferromagnetic material to operate. For these reasons, a naval vessel whose hull and structure are made completely from nonmagnetic material may still have a ferromagnetic field signature, although significantly reduced in amplitude.

An estimate of the ferromagnetic field of individual shipboard items can be computed once again using the prolate spheroidal shell model. A worse case condition will be assumed for this example consisting of a completely solid onboard item with a permeability constant of 500. The length-to-diameter ratio will be fixed at 4 while the Earth's longitudinal inducing field at 55,000 nT. The off-board signatures 20 m directly below the item were computed for various lengths and are plotted in Figures 2.7 and 2.8 for the vertical and longitudinal components, respectively. Although a single item's magnetic field amplitude is much less than that of an entire ferromagnetic hull, it is not zero. In addition, nonmagnetic hull ships can be equipped with many magnetic items whose

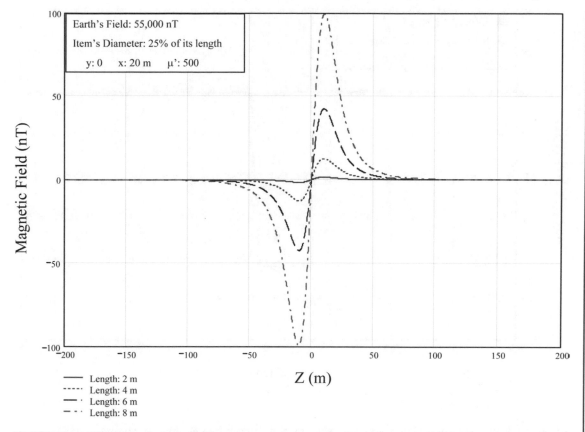

FIGURE 2.7: Vertical magnetic field signatures produced by the induced longitudinal magnetization of small prolate spheroids of varying sizes.

magnetic fields can add together to produce a sizable net off-board ferromagnetic signature. In fact, minesweepers can have more than 100 permanently installed magnetic equipment items. If these items or their parts cannot be replaced with nonmagnetic functionally equivalent components and the ship's overall signature is still higher than requirements, then active field cancellation techniques (degaussing systems) must be used to reduce the signature to the desired level.

Ships and submarines constructed entirely from magnetic steel have two ferromagnetic source components, induced and permanent magnetization, where the latter can be reduced using a passive magnetic silencing technique. The permanent or residual magnetization is the result of magnetic domains within the material that remain fixed in their orientation as the external inducing field changes. The plot of a material's magnetization as a function of an externally applied magnetic field forms a highly nonlinear curve called a *hysteresis curve* [2].

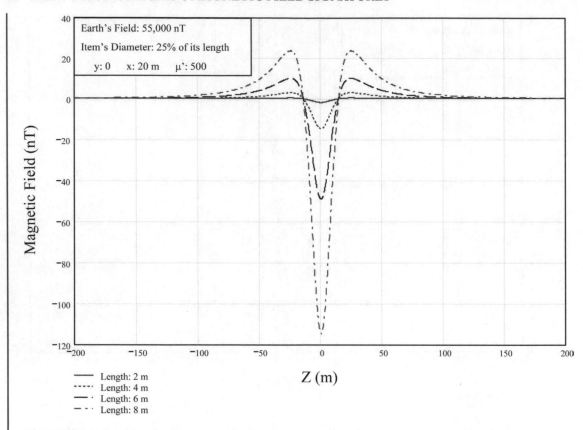

FIGURE 2.8: Longitudinal magnetic field signatures produced by the induced longitudinal magnetization of small prolate spheroids of varying sizes.

When a magnetic material is subjected to mechanical stress, high temperatures, or large fields, the magnetic domains reorient themselves, resulting in a change in the vessel's permanent magnetization. Due to manufacturing processes, surface ships and submarines typically leave the construction yard with a large permanent magnetization. Also, submarines that are subjected to significant hull stresses during deep dives can accumulate a sizable permanent magnetization, which is simply called *perm*.

Demagnetization techniques can be used to reduce the perm of a surface ship or submarine. The process of demagnetizing a naval vessel is called *deperming* and is conducted at *magnetic treatment facilities* owned and operated by naval establishments. A photograph of the USS *Higins* (DDG-76) being depermed at the San Diego Naval Magnetic Treatment Facility, San Diego, CA, is shown in Figure 2.9, whereas in Figure 2.10, the USS *Jimmy Carter* (SSN-23) is moored inside the magnetic silencing facility at the Naval Base Kitsap Bangor, Bangor, WA [11]. The San Diego

FIGURE 2.9: The USS *Higins* (DDG-76) being depermed at a close-wrap magnetic treatment facility.

station is called a *close-wrap* treatment facility since the deperming cable is wrapped close around the ship's hull as shown in the photograph, while the Bangor site is called a *drive-in* facility for obvious reasons. On the seafloor beneath and around each treatment slip is an array of fluxgate magnetometers used to monitor and control the deperming process. In addition, this sensor system measures and records the final signature of the depermed vessel to determine if it meets magnetic silencing specifications. All of the treatment facility's systems are calibrated just before the surface ship or submarine enters the slip.

The process of deperming a naval vessel is complicated by the nonlinear hysteresis of its thick steel hull. When a ship or submarine first arrives at the treatment slip, its longitudinal and athwart-ship perm could be pointing in either the positive or negative direction depending on the *magnetic history* of the vessel. Magnetic history refers to the sequence of changes in a naval vessel's perma-nent magnetization usually caused by numerous cycles of mechanical stress. The amplitude and direction of the perm change is governed primarily by the amount, distribution, and type of stress

FIGURE 2.10: The USS *Jimmy Carter* (SSN-23) being depermed at a drive-in magnetic treatment facility.

(compression, tensile, or torsion) being applied to the vessel's magnetic material, the magnitude and direction of the Earth's field that was present during each stress cycle, and the starting point on the hysteresis curve. As can be imagined, it is virtually impossible to keep track of these changes in perm.

The polarity of the vertical component of a ship's perm is somewhat more predictable than the horizontal components. Vessels that sail primarily in the northern magnetic hemisphere typically accumulate a net positive perm (downward pointing) since they experience a consistent downward pointing Earth's field in northern latitudes during stress cycles, whereas those that spend the bulk of their time in the southern hemisphere tend toward negative pointing vertical perm vectors. Ships that cross the magnetic equator could have either a positive or negative vertical perm on arrival at the treatment facility.

In general, the objective of ship deperming is to minimize its longitudinal, athwartship, and vertical perm components. However, there are circumstances under which a vertical perm is deliberately imparted to the vessel. If it is to operate in a small zone about specific magnetic latitudes, then a vertical perm may be purposely setup in the hull that will cancel the induced vertical magnetization in its zone of operation. This deperming technique is called *flash-deperming*. Also, vessels have been deliberately magnetized in the vertical direction to produce a perm distribution that is consistent between hulls in an effort to simplify degaussing coil designs.

A surface ship or submarine is depermed by first applying large positive and negative magnetic fields in a cyclic fashion along its longitudinal axis, and then slowly reducing their amplitude. The large cyclic deperming fields are generated by passing thousands of amperes through the close-wrap or drive-in solenoid. If the cyclic magnetic fields, called *shots*, are applied within a zero background field, the perm vectors will all be minimized. If a direct current (DC) bias field is placed on the vessel during the shots, then a perm will be setup in the direction of the bias. In this introduction to the magnetic silencing technique of deperming, it will be assumed that all perm vector components are to be minimized using a zero bias.

The ship deperming process is easiest to explain graphically using a hysteresis curve. In this example, a ship arrives at the deperming facility with a large negative permanent longitudinal magnetization (PLM) that is to be reduced as close as possible to zero. First, the Earth's magnetic field is cancelled out so that the net bias field on the vessel is near zero. This is accomplished by injecting a small bias current into the longitudinal close-wrap or drive-in solenoid, sometimes called the *x* loop, to zero out that component of the background field. Since deperming facilities are built in a magnetic north–south direction, no athwartship bias is required. Cancellation of the vertical component of the Earth's field is accomplished with a large horizontal loop, called a *z* loop, that is installed either on the seafloor or incorporated into the facility's structure. This magnetic state of the ship is labeled as point #1 on the hysteresis curve in Figure 2.11.

With the background field on the surface ship or submarine zeroed out by the facility, the deperming process can begin. In this example, the first magnetic deperming shot will be generated by passing several thousand amperes through the *x* loop so as to produce a large positive magnetic field along the vessel's longitudinal axis, while accurately maintaining the smaller bias current needed to cancel the Earth's field. When this large *top shot* is applied, the ship's magnetization will move to the point labeled #2 in Figure 2.11, which should ideally be near the positive saturation level for the ship's steel. A top-shot current and field is typically held on for about 1 minute to ensure that all eddy currents have died out and the deperming field has fully penetrated the hull. After applying such large current for this amount of time, the *x*-loop cables must be allowed to cool for several minutes before starting the next top shot.

The second top shot in this example will be in the negative direction, opposite to the polarity of the first. The second shot will move the ship's magnetization from point #2 to the point labeled #3 in Figure 2.11 near the negative saturation level. After another cable cool-down period, the third top shot will bring the magnetization to point #4, which ideally will be in the neighborhood of point #2. It should be obvious that at this stage in the deperming process, the vessel's magnetization

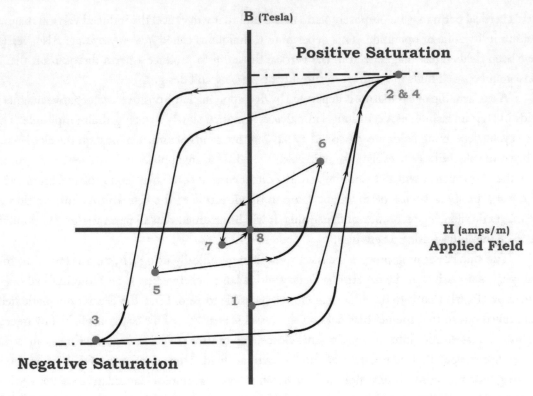

FIGURE 2.11: Example hysteresis pattern produced by a ship undergoing the deperming process.

will end up at point #4 regardless of where its starting point (#1) happens be located. Even if the initial perm were positive, the three specified top shots should still bring the magnetization back to the vicinity of point #4.

After completing the top shots, the polarity of the deperming field is alternately cycled while reducing its amplitude. In this example, subsequent lower-level deperming shots will move the ship's magnetization sequentially from point #4 to #5, to #6, to #7, and then ideally to the origin at point #8. There are many practical reasons for not achieving a zero perm signature at the end of the deperming cycle, some of which are:

1. The bias field was not set correctly at the beginning of the deperming.
2. The bias field drifted during the process.
3. Insufficient number of top shots.
4. The amplitudes of the cyclic deperming shots were reduced too quickly.

If the final magnetic state of the surface ship or submarine does not meet perm signature requirements the long deperming process must be repeated. The surface ship and submarine deperming process has been studied in depth using a scale model and is reported by Baynes et al. [12].

2.2   PASSIVE REDUCTION OF ROLL-INDUCED EDDY CURRENT SIGNATURES

Under some conditions, the uncompensated roll-induced eddy current magnetic field signatures of naval vessels can be comparable to their ferromagnetic. This makes eddy current signatures the second most important source contributing to a ship's underwater magnetic field. Eddy currents are generated in any electrically conducting material found onboard a ship as it rotates within the Earth's magnetic field. This process is similar to what occurs inside an electric generator as its rotating windings cut through the static magnetic flux lines established inside it. Shipborne eddy currents produce their own magnetic fields that can modify and add to a ship's ferromagnetic signatures that lay in the ultralow frequency (ULF) passband of an influence mine.

Ships made from aluminum, stainless steel, or titanium will all have eddy currents induced in them as they rotate, although they are nonmagnetic. A vessel does not have to be ferromagnetic to support eddy currents and their associated underwater magnetic signatures. From Faraday's law, the eddy current density $\vec{J}$ is given by:

$$\vec{J} = \sigma \left(\vec{v} \times \vec{B}_e \right)$$

(2.2)

where σ is the conductivity of the ship material, $\vec{v}$ is its velocity vector, and $\vec{B}_e$ is the Earth's static magnetic field vector. Although in principal eddy currents are produced when a vessel pitches or changes heading, these components are generally much smaller in comparison to the roll-induced currents being discussed here.

Some of the important characteristics of roll-induced eddy current signatures can be explained using a conducting loop of wire as a model. Electric currents are induced in the wire loop shown in Figure 2.12a if it is rotated within the Earth's magnetic field, or if the loop is held stationary and an external alternating current (AC) magnetic field is imparted to it. An equivalent circuit for the wire loop is drawn in Figure 2.12b. The source in the circuit represents the voltage v_e induced in the loop by the enclosed time-varying magnetic flux and is given here by:

$$v_e = -\frac{d\Phi}{dt}$$

(2.3)

a.) Induction loop   b.) Equivalent circuit

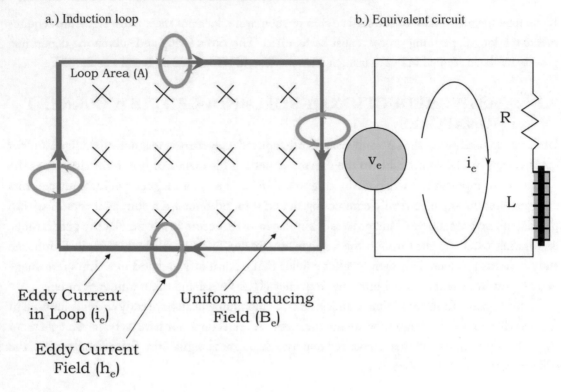

FIGURE 2.12: Equivalent circuit representation of roll-induced eddy currents.

where

$$\Phi \propto AB_e\theta_{max}e^{j\omega t}$$
$$\omega = 2\pi f$$

and A is the area of the loop, B_e is the magnitude of the Earth's field, θ_{max} is the maximum of a small roll angle, and f is the roll frequency. From the analysis of the simple circuit, the eddy current, i_e, in the wire and its field, h_e, are proportional to:

$$h_e \propto i_e \propto \frac{-j\omega AB_e\theta_{max}e^{j\omega t}}{R+j\omega L} \tag{2.4}$$

where R and L are the equivalent resistance and inductance of the circuit, respectively. It should be noted that eddy current fields are proportional not only to the ship's maximum roll angle but also to its roll frequency. Equation (2.4) shows that eddy currents and their fields will have both real and quadrature components, a characteristic that is important when attempting to actively compensate it.

The impact of material selection on the eddy current signatures of naval vessels can be significant and will be demonstrated with a simple two-dimensional example. In this case, a long cylindrical conducting shell with an inner and outer radius of a and b, respectively, will be subjected to a radial AC magnetic inducing field of amplitude, B_e, and frequency f. (The example's geometry and coordinate system are shown in Figure 2.13.) The shell material will represent a ship's hull with

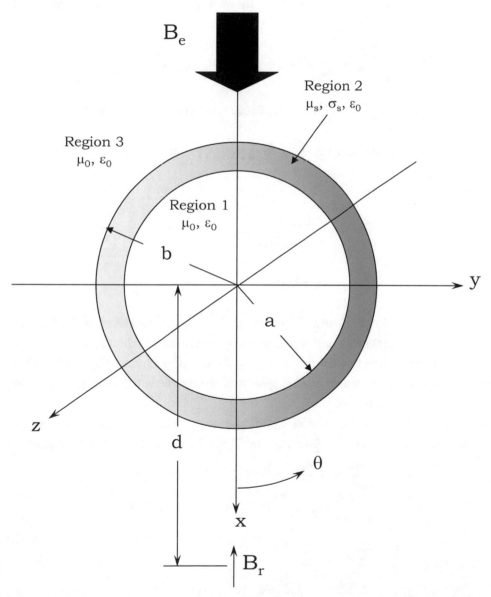

FIGURE 2.13: Coordinate system for the two-dimensional roll-induced eddy current problem.

a conductivity of σ_s and magnetic permeability of μ_s, while its interior and exterior will be taken as free space. The vertical component of magnetic field will be computed directly below the shell's center at a depth, d.

The two-dimensional roll-induced eddy current shell problem can be formulated using a quasi-static approach. Due to the symmetry of the problem, only the longitudinal component of the magnetic vector potential, $\vec{A}$, is needed. Also, the magnetic flux density $\vec{B}$ is given by $\vec{B} = \nabla \times \vec{A}$. Assuming a $e^{j\omega t}$ time variation as in Equation (2.3), the general vector potential solutions suitable for the three regions of Figure 2.13 are given as:

$$A_{z_1} = \sum_{n=1}^{\infty} r^n \left(A_n \cos(n\theta) + B_n \sin(n\theta) \right) \tag{2.5a}$$

$$A_{z_2} = \sum_{n=1}^{\infty} \left(C_n I_n(\gamma_s r) + D_n K_n(\gamma_s r) \right) \left(E_n \cos(n\theta) + F_n \sin(n\theta) \right) \tag{2.5b}$$

$$A_{z_3} = B_i r \sin(\theta) + \sum_{n=1}^{\infty} r^{-n} \left(G_n \cos(n\theta) + H_n \sin(n\theta) \right) \tag{2.5c}$$

where

$$\gamma_s = \sqrt{j\omega \mu_s \sigma_s}$$

$$B_i = B_e \theta_{max}$$

and r is the radial coordinate, I_n and K_n are the modified Bessel functions of the first and second type, respectively, A_n through H_n are constants to be determined by boundary conditions, and all other parameters have been defined previously. As discussed by Holmes [2], the boundary conditions are the continuity of the normal component of the magnetic flux density and the tangential components of the field intensity.

A system of equations can be established using the boundary conditions to solve for the unknown constants in Equations (2.5a), (2.5b), and (2.5c). In terms of the vector potentials they are:

$$\frac{\partial A_{z_1}}{\partial \theta} = \frac{\partial A_{z_2}}{\partial \theta}$$
$$\frac{1}{\mu_0} \frac{\partial A_{z_1}}{\partial r} = \frac{1}{\mu_s} \frac{\partial A_{z_2}}{\partial r} \qquad \text{at } r = a \tag{2.6a}$$

$$\frac{\partial A_{z_3}}{\partial \theta} = \frac{\partial A_{z_2}}{\partial \theta} \qquad \text{at } r = b \tag{2.6b}$$
$$\frac{1}{\mu_0} \frac{\partial A_{z_3}}{\partial r} = \frac{1}{\mu_s} \frac{\partial A_{z_2}}{\partial r}$$

It should be pointed out that only the $n = 1$ terms in Equations (2.5a), (2.5b), and (2.5c) can be used to match the boundary conditions established by the AC background field. Substituting the $n = 1$ terms of Equation (2.5a), (2.5b), and (2.5c) into Equations (2.6a) and (2.6b) gives the system of equations:

$$aB_1 = C_1 I_1 (\gamma_s a) + D_1 K_1 (\gamma_s a) \tag{2.7a}$$

$$\frac{1}{\mu_0} B_1 = \frac{\gamma_s}{\mu_s} \left(C_1 I_1' (\gamma_s a) + D_1 K_1' (\gamma_s a) \right) \tag{2.7b}$$

$$B_i + \frac{H_1}{b^2} = \frac{1}{b} \left(C_1 I_1 (\gamma_s b) + D_1 K_1 (\gamma_s b) \right) \tag{2.7c}$$

$$\frac{1}{\mu_0} B_i - \frac{1}{\mu_0} \frac{H_1}{b^2} = \frac{\gamma_s}{\mu_s} \left(C_1 I_1' (\gamma_s b) + D_1 K_1' (\gamma_s b) \right) \tag{2.7d}$$

The terms A_1, F_1, and G_1 are not needed to satisfy the boundary conditions, and F_1 has been incorporated into C_1 and D_1. Since the field in region 3 is of interest in this example, only the H_1 constant is needed here, which can be written as:

$$H_1 = -b^2 B_i \frac{a (\eta_1 + \mu_s \eta_2) - \mu_s (\eta_3 + \mu_s \eta_4)}{a (\eta_1 - \mu_s \eta_2) - \mu_s (\eta_3 - \mu_s \eta_4)} \tag{2.8}$$

where:

$$\eta_1 = b\gamma_s^2 \left(K_1' (\gamma_s a) I_1' (\gamma_s b) - I_1' (\gamma_s a) K_1' (\gamma_s b) \right)$$

$$\eta_2 = \gamma_s \left(I_1' (\gamma_s a) K_1 (\gamma_s b) - K_1' (\gamma_s a) I_1 (\gamma_s b) \right)$$

$$\eta_3 = b\gamma_s^2 \left(K_1 (\gamma_s a) I_1' (\gamma_s b) - I_1 (\gamma_s a) K_1' (\gamma_s b) \right)$$

$$\eta_4 = I_1 (\gamma_s a) K_1 (\gamma_s b) - K_1 (\gamma_s a) I_1 (\gamma_s b).$$

The equations for the off-board roll-induced eddy current signatures become:

$$B_r = \frac{H_1}{r^2} \cos(\theta) \tag{2.9}$$

$$B_\theta = \frac{H_1}{r^2} \sin(\theta) \tag{2.10}$$

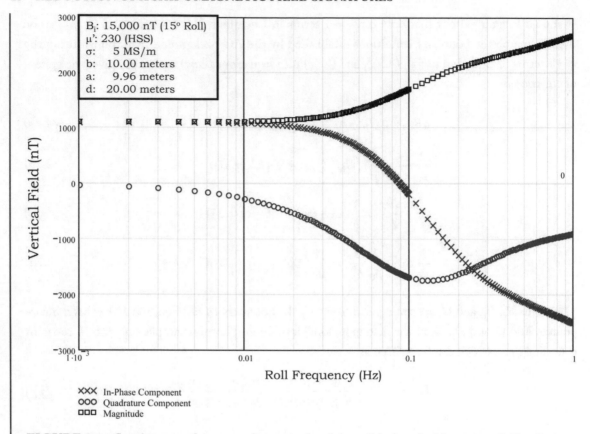

FIGURE 2.14: In-phase, quadrature, and magnitude of the roll-induced eddy current fields of a two-dimensional steel cylindrical shell.

If the external inducing field is expressed in nT, so will the fields given by Equations (2.9) and (2.10).

Roll-induced eddy current signatures have two temporal components: a real constituent that is either in-phase or 180° out-of-phase with the inducing field and a quadrature component that is 90° out-of-phase. To demonstrate this, Equation (2.9) was evaluated for HSS at a depth of 20 m, just below the cylinder's centerline corresponding to $\theta = 0°$. The magnitude of the field and its in-phase and quadrature components are plotted in Figure 2.14 as a function of roll frequency. All other parameters for this example are listed in the figure.

Actively compensating both the in-phase and quadrature components for all the roll frequencies of interest can be difficult. As suggested by Equation (2.4) and demonstrated in Figure 2.14, the in-phase component of the roll signature of an HSS-hull vessel reduces to its static ferromagnetic field at long roll periods and becomes more diamagnetic as the frequency increases. The quadrature component starts at zero, increases in the negative direction with frequency, and then returns back to zero. It should be pointed out that a naval influence mine detects the vector sum (magnitude) of the in-phase and quadrature components.

The roll-induced magnetic field signature of an aluminum hull ship can still be significant although it is nonmagnetic. If Equation (2.9) is recomputed using the constituent parameters for aluminum, its in-phase, quadrature, and vector magnitude shown in Figure 2.15 are produced at a depth of 20 m. All other parameters are the same as the HSS example. Although aluminum is nonmagnetic, a diamagnetic in-phase component is generated as the roll frequency increases from zero

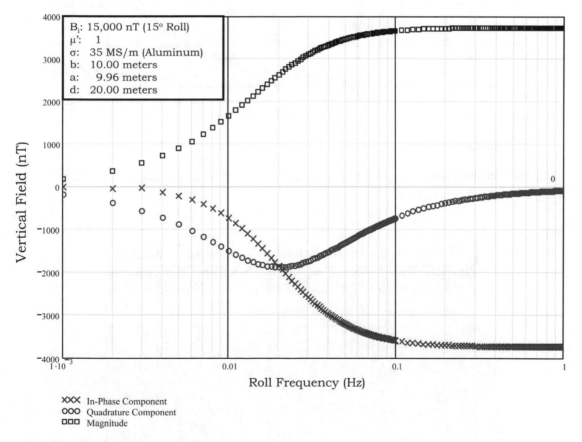

FIGURE 2.15: In-phase, quadrature, and magnitude of the roll-induced eddy current fields of a two-dimensional aluminum cylindrical shell.

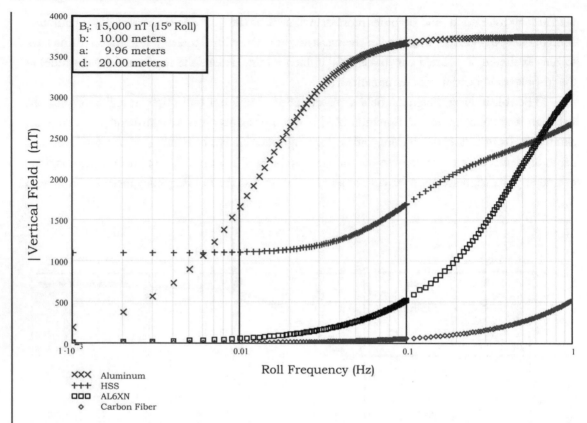

FIGURE 2.16: Magnitude of the roll-induced eddy current fields of a two-dimensional cylindrical shell of varying material composition.

TABLE 2.2: Electric conductivity of ship construction materials	
MATERIAL	σ **(MS/m)**
Pure aluminum	35
High-strength steel	5
Pure titanium	2
AL6XN stainless steel	1
Carbon	0.1
Fiberglas	Nil

and reaches its maximum negative value rather quickly. The quadrature field's frequency response has the same shape as HSS but peaks at a much lower roll frequency.

The underwater roll-induced eddy current magnetic field signature for different ship construction materials can be evaluated and compared using Equation (2.9). The electrical conductivities of several hull materials are given in Table 2.2 in the units of MS/m. The magnitude of the vertical magnetic field component as a function of roll frequency can be compared in Figure 2.16 for the various materials. As shown by the data, aluminum hull vessels can still have a significant magnetic field due to the roll-induced eddy current source although the material is nonmagnetic. This is important when considering the sea mine susceptibility of fast attack boats used by special operations forces.

A hull material that is a low signature alternative to HSS armor plating that could be used on heavy combatants is AL6XN stainless steel. Not only is it nonmagnetic, but it also has a

FIGURE 2.17: Eddy current flow induced in a conducting hulled ship due to its roll in the Earth's magnetic field.

significantly lower eddy current field in the vicinity of 0.1 Hz, the natural roll frequency of large surface ships. As shown, hulls constructed from carbon fiber composites would have an extremely low, but nonzero, eddy field. However, vessels built entirely from fiberglass or wood have an unmeasurable roll-induced eddy current field.

Roll-induced eddy currents produce magnetic dipolar type sources that are primarily oriented in the vertical and athwartship directions. The general flow pattern of these currents on a conducting hull ship has been drawn in Figure 2.17. The off-board magnetic fields generated by the eddy current sources have the same general characteristics as the uncompensated vertical and athwartship ferromagnetic signatures [4]. This means that the field components from both of these sources can, in principle, be cancelled with degaussing coils. This active signature reduction technique will be discussed in the next chapter.

2.3   PASSIVE REDUCTION OF CORROSION-RELATED MAGNETIC FIELD SIGNATURES

The least known of the major shipboard sources of magnetic field are corrosion-related electric currents that flow in and around a surface ship or submarine's hull. When two or more electrically conducting materials of different electrochemical potentials are connected together and immersed in seawater, a battery is formed. In this case, an electric current (conventional current indicating positive charge flow) will leave the more electrochemically negative material called the *anode*, flow in the conducting seawater to the more positive called the *cathode*, and then return back to the anode through their connection point. (Electric current flows in seawater by a different mechanism than it does in a metallic conductor as discussed by Holmes [4].) In this *freely corroding* state, the anode material will rust.

Predicting the precise flow of corrosion current between different materials is an involved task. In seawater, typical ship construction materials have nonlinear polarization curves that relate their electrochemical potentials to corrosion current density. The problem is complicated by the numerous chemical reactions that occur, which are dependent on the conditions of protective hull paints (coatings), the area exposed to seawater, the velocity of the water flow across the material, etc. Typically, boundary element techniques are used to model a ship's corrosion currents and associated signatures [2].

The greater the difference in electrochemical potentials between the anode and cathode the greater will be the electric current flow in the corrosion circuit. The average open-circuit electrochemical potential of several ship materials, measured in seawater relative to a silver–silver chloride reference, is listed in Table 2.3. As the table shows, a vessel's HSS hull will be anodic with respect to a nickel–aluminum–bronze (NAB) propeller, if allowed to freely corrode, and will have an initial potential difference of approximately 420 mV. To prevent corrosion, a vessel's ICCP system would

HULL MATERIAL	ϕ (mV)
Aluminum alloys	−800
High-strength steel	−650
Nickel–aluminum–bronze (NAB) propeller alloy	−230
Titanium alloys	~0
AL6XN stainless steel	~0
Graphite carbon	+25

TABLE 2.3: Electrochemical potential of ship construction materials

typically be set so that its controlling reference cell maintains a potential relative to the hull that is more negative than the most electrochemically active material. This raises the current flow through the propeller, shaft, and hull over that of the freely corroding state and could produce a larger magnetic field signature.

As implied by Table 2.3, an unprotected section of an aluminum hull ship will have a larger corrosion-related current flow and magnetic field than an equivalent HSS vessel. The potential difference between aluminum and NAB is 570 mV, almost three times greater than a hull built from HSS. This is an example of why care must be taken when designing signature reduction systems. The designer should avoid decreasing one source's amplitude at the expense of increasing another's.

Surface ship and submarine hulls constructed from either AL6XN-type stainless steel or titanium would be expected to have a lower corrosion-related magnetic (CRM) field signature. The potential difference between a hull constructed from AL6XN stainless and a NAB propeller is only 230 mV, almost twice as small as an HSS hull. In addition, an ICCP system for a stainless steel hull could be set much lower than HSS or aluminum, possibly as low as -430 mV. It should be noted that a ship or boat built from a carbon fiber composite may still have corrosion issues. In any case, naval vessels constructed from the materials listed at the bottom of Table 2.3 would have lower CRM fields.

Since ship propellers are generally on their aft end, the major corrosion currents will flow in the forward-aft direction. At the simplest level, a vessel's corrosion current source can be represented by an extended electric dipole source aligned in the longitudinal direction, which will be taken here as the z axis of a coordinate system. The magnetic field produced by this source circulates around the hull according to the right-hand rule (see Figure 2.18). Assuming the corrosion current is flowing from the propeller located at $z = -L/2$, through the shaft and hull, to a point along the vessel given by $z = L/2$, the φ component of the magnetic field B_φ is given by [13]:

$$B_\varphi = \frac{\mu_0 I}{4\pi\rho} \left(\frac{z + \dfrac{L}{2}}{r_1} - \frac{z - \dfrac{L}{2}}{r_2} \right) \tag{2.11}$$

where

$$r_1 = \sqrt{\left(z + \frac{L}{2}\right)^2 + \rho^2}$$

$$r_2 = \sqrt{\left(z - \frac{L}{2}\right)^2 + \rho^2}$$

FIGURE 2.18: Magnetic field pattern around a ship produced by corrosion-related currents.

and I is the source current, L is the effective length of the dipole, and ρ is the radial coordinate. From Equation (2.11), the peak magnetic field will occur at $z = 0$, and can be computed from:

$$B_{\text{peak}} = \frac{\mu_0 IL}{4\pi\rho\sqrt{\left(\frac{L}{2}\right)^2 + \rho^2}} \qquad (2.12)$$

A simple example can be used to demonstrate the relationship between the corrosion current source and the magnetic field it produces.

As given by Equation (2.12), the peak CRM field of a naval vessel is directly proportional to the current magnitude. The peak magnetic fields under the extended electric dipole source at a depth of $\rho = 20$ m were computed for a number of source currents and dipole lengths and are plotted in Figure 2.19. If the corrosion or cathodic protection current amplitudes can be decreased through the proper selection of hull material, then its off-board magnetic field can be reduced by the same

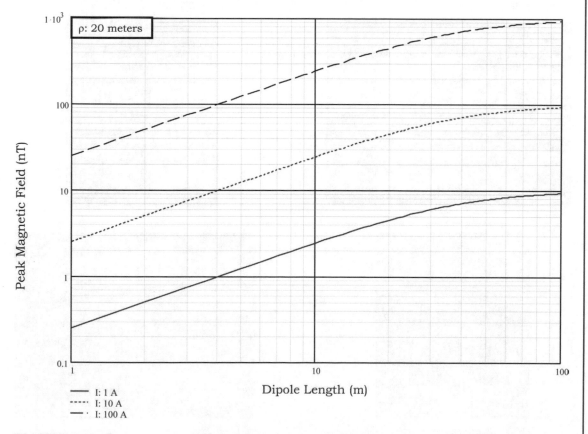

FIGURE 2.19: Peak magnetic fields produced by electric dipoles of different current strengths as a function of their lengths.

amount. Also, if the anodic current source is moved closer to the propeller cathode, then the CRM signature can still be reduced according to Equation (2.12) even if the current magnitude is not. Shorter dipole lengths are the primary objective of an active signature cancellation system, which is to be discussed in the next chapter.

Since the NAB propeller is the major sink of current on a surface ship or submarine, any increase in the electrical resistance of its corrosion circuit will decrease the current flow through it. One method to reduce the corrosion current into a propeller is to coat it with high-resistance paint. The four-step coating process depicted in Figure 2.20 is an example of this technique. The air temperature and humidity must be in the proper range during the application of the paint layers or the coatings may not adhere properly. Failure of the propeller coatings usually begin near the high-speed blade tips, and progress inward. However, good paint adhesion can be achieved if care is taken during its application.

a.) Base Coat

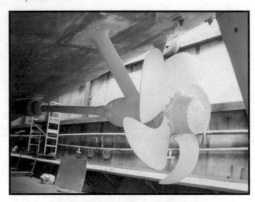

b.) Mid Coat

c.) Antifouling Coat

d.) Top Coat

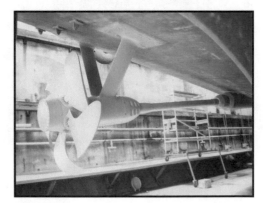

FIGURE 2.20: Four-step coating (paint) being applied to a nickel–aluminum–bronze propeller.

A major drawback of using propeller paints is the necessity of removing them during structural inspections. Paints fill into microcracks within the propeller material, preventing them from being detected. Abrasive removal of the coatings is required before inspection, and then the propeller must be repainted.

Corrosion currents are also an important source of alternating magnetic field signatures. As corrosion currents flow up a propeller's shaft and through its bearings on its way back to the hull, any variability in the shaft-to-bearing resistance will cause the corrosion currents to be modulated. The impulsive nature of the modulated currents produces magnetic fields at the shaft's fundamental rotation frequency plus harmonics [14].

Shaft grounding systems are typically used to reduce the shaft's modulation of the corrosion currents along with its associated magnetic field. A passive shaft grounding system consists of a slip ring attached to the shaft, and a silver-tipped carbon brush that rides on the slip ring under spring tension, which is grounded by cable to the vessel's hull (see Figure 2.21a). The purpose of this grounding system is to provide a low-impedance path to the hull that will bypass the bearing modulation mechanisms, effectively shorting them out. However, if any grease or dirt finds its way under the carbon brush, the resistance of the ground circuit quickly rises, making it an ineffective ground and signature reduction device.

An active shaft grounding (ASG) system is more reliable than a single passive brush. As shown in Figure 2.21b, an ASG has two independent slip rings and brushes attached to the shaft. The first brush measures the shaft-to-hull potential through a high-input impedance electronic

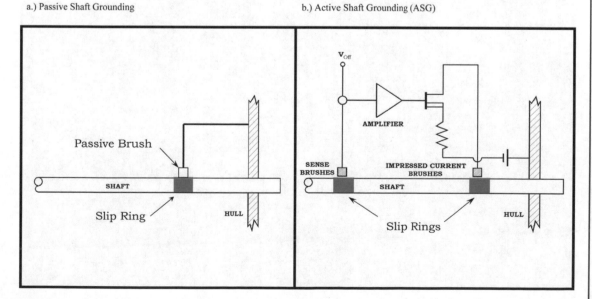

FIGURE 2.21: Circuit diagram for a passive and active shaft grounding system.

amplifier that makes the measurement somewhat insensitive to variations in brush-to-slip ring impedance. The ASG supplies current to the shaft by way of a second slip ring and brush so that the shaft-to-hull voltage is minimized. This condition also effectively shorts out the bearings and the modulation mechanism. If the impedance of the second brush-to-slip ring increases, the ASG output voltage increases (within its capacity) to a level necessary to inject the required current. The insensitivity of the ASG to brush impedance changes makes its performance superior to the passive shaft grounding system in reducing a naval vessel's alternating magnetic field signature [14, 15].

The ASG system is considered a passive signature reduction system although it is composed of active electronics. The objective of the ASG is to reduce or eliminate the shaft-modulated corrosion current, which is the source of the ship's alternating magnetic field. The ASG system cancels the signature's source directly as opposed to compensating its field.

2.4   PASSIVE REDUCTION OF STRAY FIELD SIGNATURES

Stray field signatures can be produced by any current-carrying electric circuit found onboard a ship. The larger of the stray fields are produced by the vessel's electromechanical machinery and power

FIGURE 2.22: Stray field sources found onboard naval vessels.

distribution system. High-power electric generators, motors, switchgear, breakers, and the distribution cables that interconnect them can emit both DC and AC fields (Figure 2.22). Ironically, a
ferromagnetic steel hull shields the internal stray field sources to some degree, especially at higher
frequencies. However, even a naval vessel's relatively thick magnetic hull does not provide significant field attenuation near DC.

   Shielding effectiveness of power distribution cables inside a ship or submarine's hull can be
estimated with a two-dimensional cylindrical shell model. The model's coordinate system and geometry are given in Figure 2.23. A distribution cable carrying a current, I, is located at $(0, -c)$ inside
an infinitely long cylindrical magnetic shield of conductivity, σ_s, and permeability, μ_s, and with an
inside radius, a, and outside radius, b. The return distribution cable is located at $(0, c)$ and has a

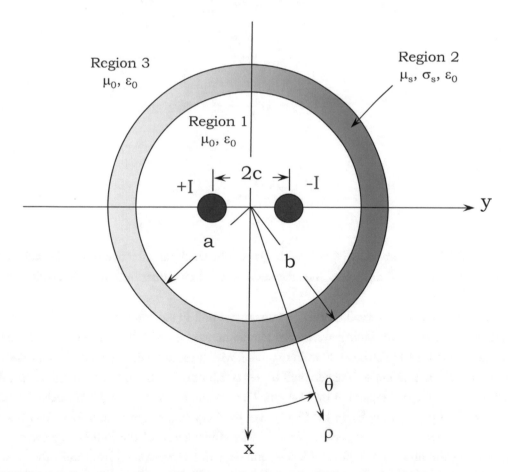

FIGURE 2.23: Coordinate system for the two-dimensional cylindrical shell shielding example.

current magnitude of $-I$. The expression for the shielding effectiveness S_{H} can be found in the work of Hasselgren and Luomi [16], and is given by:

$$S_{\mathrm{H}} = 20 \log \left(\frac{|\nabla X_0|}{|\nabla X_{\mathrm{s}}|} \right) \qquad (2.13)$$

where

$$X_{\mathrm{s}} = -\frac{I}{\pi} \sum_{n=1,\,3,\,5,\ldots}^{\infty} \frac{1}{n} \left(\frac{c}{\rho} \right)^n Q_n \sin n\varphi$$

$$Q_n = \frac{1}{\cosh\left(k_n t\right) + \frac{1}{2}\left(K_n + \frac{1}{K_n} \right) \sinh\left(k_n t\right)}$$

$$K_n = \frac{k_n b}{\mu_{\mathrm{r}} n}$$

$$k_n = \sqrt{\frac{n^2}{b^2} + \gamma_{\mathrm{s}}^2}$$

$$\gamma_{\mathrm{s}} = \sqrt{j\omega \mu_{\mathrm{s}} \sigma_{\mathrm{s}}}$$

$$\mu_{\mathrm{r}} = \frac{\mu_{\mathrm{s}}}{\mu_0}$$

$$t = b - a$$

and the unshielded case given by $X_0 = X_{\mathrm{s}}$ with $Q = 1$. The shielding effectiveness in decibels can be estimated for various hull diameters and magnetic permeability, distribution cable configurations, and current frequencies.

The importance of a naval vessel's ferromagnetic hull in shielding the magnetic fields from internal electric power distribution systems can be demonstrated with Equation (2.13). In this example, the hull will be HY80 steel 20 m in diameter, with a conductivity of 3.5 MS/m and relative permeability of 90. The hull's thickness will be set at 1.3 cm, and the internal distribution cables located in the center and separated by 11.4 cm. The shielding effectiveness in decibels at a depth of 20 m was computed using Equation (2.13) over the frequency range from 0.01 to 100 Hz, and is plotted in Figure 2.24. As expected, the shielding effectiveness of the hull is negligible at low frequencies on the order of 0.1 Hz and below, and does not become significant until the current's frequency approaches 10 Hz. It should be noted that some high-power permanent magnetic motor designs are controlled with pulse width modulated current, which, when operated at slow speeds, have important frequency components in the hull's low shielding region.

Proper up-front design of a power system and its distribution cables can reduce a significant portion of its stray fields cheaply and with low impact on the ship. As an example, consider the two-conductor power cables discussed above. If the two power cables were divided into six or eight separate conducting cables, they can be configured so as to significantly reduce their stray magnetic field signatures. To demonstrate this, a simple two-dimensional formulation of the magnetic field from a single, infinitely long current-carrying conductor can be used, which is given by:

$$B_\varphi = \frac{\mu_0 I}{2\pi\sqrt{(x-x_0)^2 + (y-y_0)^2}}$$

(2.14)

where x_0 and y_0 are the coordinates of an individual cable. The peak magnetic fields from each cable in a two-, six-, and eight-conductor distribution system were computed with Equation (2.14) and

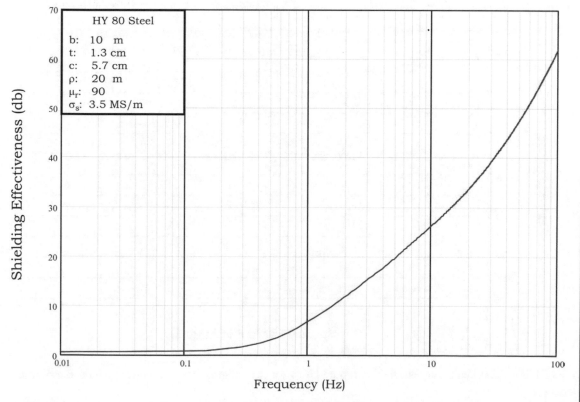

FIGURE 2.24: Shielding effectiveness of a two-dimensional HSS cylindrical shell.

then summed 20 m below the center of the bundle. As plotted in Figure 2.25, the 3000-A current flow is divided evenly between the multiple cables that are separated by a variable distance, *a*, and arranged in the three configurations drawn. As demonstrated with this example, proper design of the power distribution system can reduce stray magnetic fields by more than 3 orders of magnitude. Such a design change would be difficult and expensive to implement after the vessel has been constructed.

The magnitude and importance of magnetic stray field signatures will increase in the near future. The U.S. Navy has committed to developing an "all-electric" ship that will use large electric motors for propulsion. Since the power supplied to electric propulsion motors could exceed 30 MW,

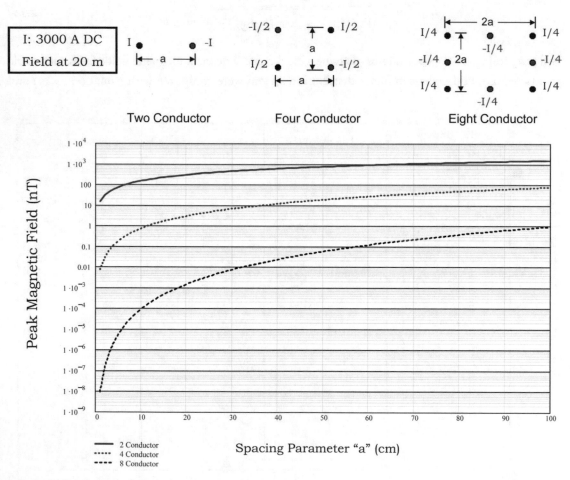

FIGURE 2.25: Peak magnetic field computed for various configurations of a power system's distribution cables.

very high voltages and, more importantly, very large currents would be flowing inside the ship's power system. The problem would be exacerbated if the motors are mounted exterior to the ferrous hull, where no shielding at all would be present. Both the DC and AC stray field signature components must be combined with the other three sources in assessing a vessel's true susceptibility to magnetic field detection.

REFERENCES

[1] J. Pike (2007, Jan). Littoral Combat Ship Specifications. Military. GlobalSecurity.org. Alexandria, VA. [Online]. Available: http://www.globalsecurity.org/military/systems/ship/lcs-specs.htm.

[2] J. J. Holmes, *Modeling a Ship's Ferromagnetic Signatures*, 1st edn. Morgan & Claypool Publishers, San Rafael, CA, 2007. doi:10.2200/S00092ED1V01Y200706CEM016

[3] M. Fogiel, *The Electromagnetics Problem Solver*. Research and Education Association, Piscataway, NJ, 2000, pp. B-1–B-3.

[4] J. J. Holmes, *Exploitation of a Ship's Magnetic Field Signatures*, 1st edn. Morgan & Claypool Publishers, San Rafael, CA, 2006. doi:10.2200/S00034ED1V01Y200605CEM009

[5] M. Fogiel, *The Electromagnetics Problem Solver*. Research and Education Association, Piscataway, NJ, 2000, p. A-2.

[6] M. Fogiel (2007, Jan). Magnetic Effects of Stainless Steel. Australian Stainless Steel Development Association. Brisbane, Australia. [Online]. Available: http://www.assda.asn.au/asp/index.asp?pgid=18045.

[7] M. Fogiel (2007, Jan). AL-6XN® Stainless Steel. Allegheny Ludlum Corporation. Pittsburgh, PA. [Online]. Available: http://www.alleghenyludlum.com/ludlum/pages/products/xq/asp/P.40/qx/product.html#.

[8] M. Fogiel (2007, Jan). How do military subs and ships avoid detection? *Nickel Mag*. Ontario, Canada. [Online]. Available: http://www.nickelinstitute.org/nickel/0999/5-0999n.shtml.

[9] M. Fogiel (2007, Jan). Alfa class submarine. Wikipedia. Wikimedia Foundation, St. Petersburg, FL. [Online]. Available: http://en.wikipedia.org/wiki/Alfa_class_submarine.

[10] M. Fogiel (2007, Jan). Visby class corvette. Wikipedia. Wikimedia Foundation, St. Petersburg, FL. [Online]. Available: http://en.wikipedia.org/wiki/Visby_class_corvette.

[11] J. McLain and M. Mohl. (2007, Jan). Jimmy Carter (SSN-23) Commissioning — Present. NavSource Online: Submarine Photo Archive. Baytown, TX. [Online]. Available: http://www.navsource.org/archives/08/080023b.htm.

[12] T. M. Baynes, G. J. Russell, and A. Bailey, Comparison of stepwise demagnetization techniques, *IEEE Trans. Mag.*, 38(4), Jul. 2002. doi:10.1109/TMAG.2002.1017767

[13] M. Fogiel, *The Electromagnetics Problem Solver*. Research and Education Association, Piscataway, NJ, 2000, p. 4-3.

[14] P. M. Holtham and I. G. Jeffrey, ELF signature control, *Proc. Undersea Defence Technol. (UDT) Conf.*, 1997.

[15] W. R. Davis. (2004). Active Shaft Grounding. W. R. Davis Engineering Ltd. Ottawa, Canada. [Online]. Available: http://www.davis-eng.on.ca/asg.htm.

[16] L. Hasselgren and J. Luomi, "Geometrical aspects of magnetic shielding at extremely low frequencies," *IEEE Trans. Electromagn. Compat.* 37(3), Aug. 1995. doi:10.1109/15.406530

• • • •

CHAPTER 3

Active Signature Compensation

3.1   DEGAUSSING SYSTEM DESIGN

Active cancellation of a surface ship or submarine's magnetic signature is achieved by deliberately generating a flux distribution whose magnitude and shape are identical to the uncompensated field, but of opposite polarity. The superposition of the uncompensated and the artificially produced compensating fields will tend to cancel, resulting in a lower net signature for the vessel. This concept is shown pictorially in Figure 3.1 for the ideal case. If the shipboard source of field is magnetic in nature such as induced and permanent magnetization, roll-induced eddy currents, or stray field sources, its active cancellation is called degaussing.

Degaussing systems were first developed by the United Kingdom to counter the German magnetic mines in World War II. Between September 1939 and January 1940, 44 British ships were sunk in the English Channel by the magnetic bottom mine threat [1]. Recovery of one of the bottom influence mines confirmed that the firing mechanism triggered on a ship's magnetic field and the race to develop signature reduction technology ensued within the British Admiralty and later the U.S. Navy. By the end of the war, more than 12,600 military and merchant ships were equipped with degaussing systems in the U.S. fleet alone.

Shipboard degaussing systems are composed of loops of electric cable that, when energized with the proper current, can cancel or reduce a ship's magnetic signature. Initially, these systems were designed only to compensate induced and permanent magnetization. As explained by Holmes [2], a ship's induced magnetization is dependent on its location within the Earth's magnetic field (latitude and longitude) and its orientation within the field (roll, pitch, and heading angle). Therefore, a degaussing system must be able to compensate the three orthogonal components of magnetization (longitudinal, athwartship, and vertical) independently from each other.

A degaussing coil designed to cancel a vessel's vertical magnetization is called an *M-coil*. The M-coil or main coil is subdivided into several smaller loops (*M-loops*), whose ampere-turns can be adjusted on an individual basis. An example of an M-coil design that was developed during World War II is shown in Figure 3.2, along with a drawing of the idealized flux distribution of the uncompensated (*undegaussed*) field. The M-coil can be used to cancel a ship's induced longitudinal magnetization (ILM) and its permanent vertical magnetization (PVM).

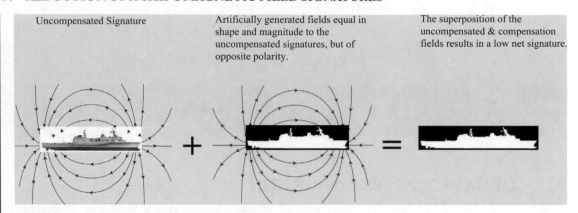

FIGURE 3.1: The process of active signature reduction systems.

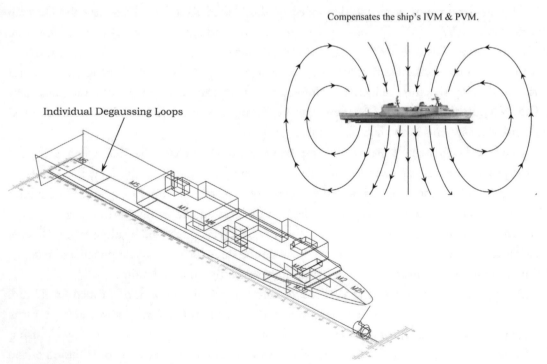

FIGURE 3.2: Old-style M-type degaussing coil designed to compensate a ship's induced vertical and permanent magnetization.

The finer signature control provided by multiple M-loops is needed in order to more precisely compensate the irregular magnetic fields produced by spatial variations in a ship's magnetization. Variations in ship magnetization are due to changes in the shape of the hull along its length, non-uniform distributions of internal magnetic structure and machinery items, and by differences in the magnetic properties of construction materials. The fields produced by nonuniform magnetization are typically represented by higher-order terms in the mathematical harmonic expansion of the undegaussed signature's source.

Compensation of the magnetic signatures produced by a ship's athwartship magnetization is accomplished with an *A-coil* or athwartship coil. A design for an A-coil is shown in Figure 3.3, along with its individual *A-loops* and undegaussed field pattern. The A-coil can compensate a naval vessel's induced athwartship magnetization (IAM) and its permanent athwartship magnetization (PAM). Two A-loops are used in tandem when the ship's beam is too large for a single A-loop located on its centerline to sufficiently compensate the off-board signature.

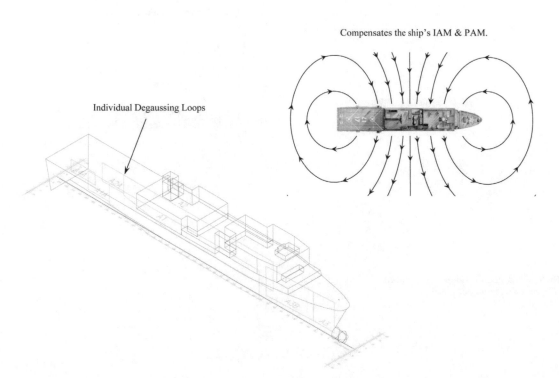

FIGURE 3.3: Old-style A-type degaussing coil designed to compensate a ship's induced athwartship and permanent magnetization.

Degaussing systems designed during World War II compensated a ship's ILM and permanent longitudinal magnetization (PLM) using a special technique that was effective against the mine threat of that time. Instead of configuring a degaussing coil to produce a longitudinal dipolar source to cancel the undegaussed magnetization in that direction, a coil configuration was developed to produce a quadrupole flux pattern. This design is called a forecastle/quarterdeck (F/Q) degaussing coil and is shown in Figure 3.4. When the F and Q coils are energized with current of opposite polarity, a flux pattern is generated beneath the vessel near its hull that tends to cancel the signature produced by the undegaussed longitudinal magnetization. The main advantage of using an F/Q-coil system over alternate designs was its ease in installation at a minimal expenditure of time and resources.

The major disadvantage of the F/Q coil is its inability to cancel ILM and PLM at larger distances from the hull. An F/Q coil forms a quadrupole whose field falls off at a more rapid rate than the undegaussed dipolar field. Because of this difference in field falloff, an F/Q coil is able to reduce

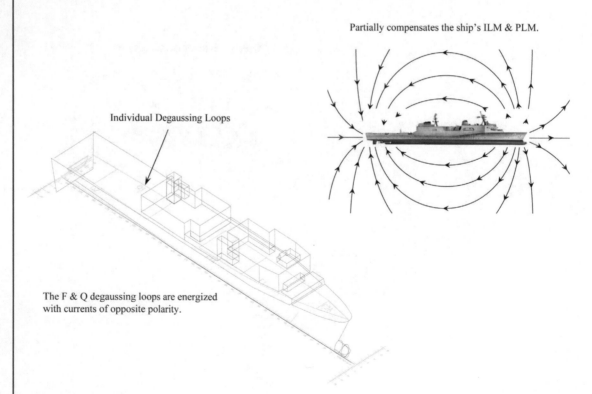

FIGURE 3.4: Old-style F/Q-type degaussing coil designed to partially compensate a ship's induced longitudinal athwartship and permanent magnetization.

a ship's ILM and PLM signatures near its hull, but with those same current settings, these field components are completely uncompensated at larger distances. However, at those larger distances where the F/Q coils lose their effectiveness, the magnitudes of the undegaussed ILM and PLM signatures are low in comparison to the relatively insensitive firing threshold of World War II mines.

Modern magnetic influence sea mines are much more sensitive than those encountered during World War II. As a result, the F/Q coil configuration is no longer adequate for compensating the ILM and PLM signatures of naval vessels. Instead, an *L-coil* system of degaussing loops has been developed as drawn in Figure 3.5. In this coil design, a number of individually controlled *L-loops* are installed along the ship's hull. An L-coil produces a dipolar flux pattern around the vessel that matches that of the undegaussed longitudinal magnetization. Therefore, the ILM and PLM signatures can be well compensated at distances both near and further away from the hull.

The development of the F/Q degaussing coil is an excellent example of *designing to meet the threat*. The pressures of World War II forced the development of a degaussing coil system that countered the magnetic mine threat that existed at the time, but used minimal resources. The L-coil configuration is more costly and time-consuming to install on a naval vessel, and its higher degree of

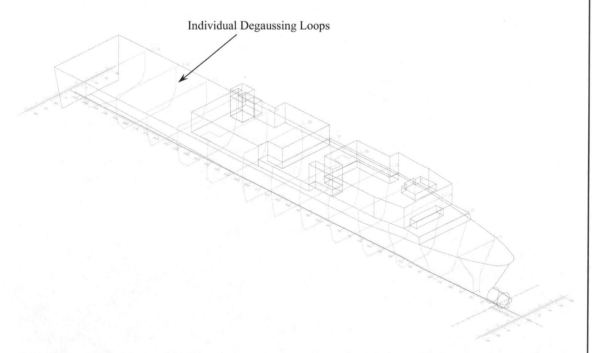

Individual Degaussing Loops

FIGURE 3.5: Modern L-type degaussing coil designed to compensate a ship's induced longitudinal and permanent magnetization.

signature compensation was not needed until the emergence of modern sea mines with much higher sensitivities to magnetic fields.

The L-coil configuration and the subdivision of the M-coil and A-coil into several loops are typical of modern degaussing system designs. An example of an advanced degaussing coil configuration is shown in Figure 3.6. Specifying the number of degaussing loops and their location within the hull of a new ship design are dependent on the shape of its hull, the magnetic properties of the material to be used in its construction, and the signature reduction level specified. Attempts have been made to establish a generalized mathematical basis for designing degaussing coils [3, 4]. For many practical reasons, the routing of degaussing coil cable throughout a ship packed with equipment and systems usually requires large deviations from their theoretical optimum paths.

In practice, the design of a degaussing system for a new ship class typically begins with a proven coil configuration pulled from historical databases that best match the requirements of the new hull. Through their implementation and use, these historical designs have been shown to be effective in meeting a specified signature reduction level while, at the same time, being practical and

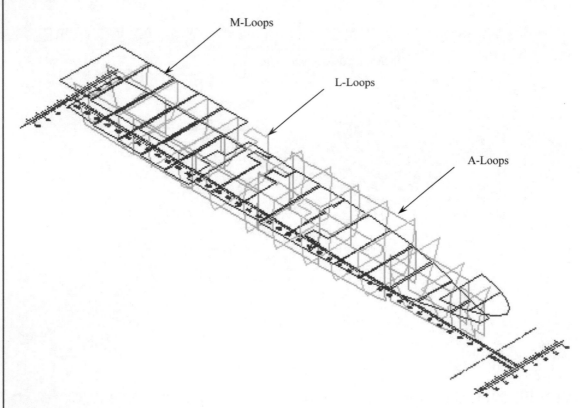

FIGURE 3.6: Advanced M-, L-, and A-type degaussing coil configuration.

cost-effective in their installation. This initial degaussing coil configuration is then modified by adding or subtracting loops and changing the geometry of existing loops to match and cancel the undegaussed magnetic signature unique to the vessel under development. Mathematical and physical scale models, which can predict the new vessel's undegaussed and compensated signatures, are tools used to optimize the degaussing coil configuration [5, 6].

3.2   DEGAUSSING COIL CALIBRATION AND CONTROL

The process of adjusting each degaussing loop's ampere-turns to minimize the ship's signature is called *calibration*. Calibrating a degaussing system requires the measurement of the vessel's undegaussed signature along with the flux pattern produced by each individual degaussing loop (loop effect). These measurements are taken at specialized shore-based degaussing calibration facilities equipped with underwater magnetic field sensors, two of which are shown in Figure 3.7.

A degaussing range (Figure 3.7a) is composed of a line of magnetic field sensors mounted on the seafloor. The vessel under calibration sails back and forth across the range while its signatures are being measured as a function of time (ranging). A tracking system is used to convert the time series data to a spatial plane of magnetic field measurements centered beneath the hull. Mathematical extrapolation models [6] are used to generate a *standard grid* of signatures that have removed from them the effects of variations in ship track and tidal changes in sensor depth that occur between successive rangings.

Degaussing ranges are installed so that the ship under calibration can sail across it on reciprocal magnetic headings. By subtracting a standard grid of field data collected when the vessel

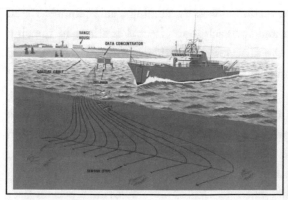

a.) Degaussing range.

b.) Fixed facility.

FIGURE 3.7: Specialized magnetic silencing facilities for measuring a ship's magnetic field signatures and calibrating its degaussing system.

was ranged on a magnetic north heading from that collected on a south, the ILM signature can be separated from the other components. (This computation actually produces two times the ILM.) Similarly, subtracting a magnetic west ranging from a east yields two times the IAM component. If the average of the magnetic north and south measurements or the average of the east and west rangings are computed, then the total perm (PLM + PVM + PAM) plus IVM signature results. Typically, degaussing ranges are not constructed with the capability to change the vertical inducing field, eliminating the possibility of using them to separate the vessel's IVM component from its total perm.

Fixed magnetic silencing facilities have been built that can separate all three induced components from the perm signature. As shown in Figure 3.7b, fixed facilities have been constructed with a set of calibration coils that surround the vessel and are used to induce a magnetization in each of the three orthogonal directions. On the seafloor beneath the ship under calibration is an array or *garden* of magnetic field sensors. The sensor garden itself is calibrated before the ship arrives to remove sensor offsets and to measure the transfer function between the facility's coil currents and their magnetic field as measured by the array. After a ship is moored in the slip, the current in each calibration coil is monitored so that its magnetic field can be subtracted from the underwater array, leaving only the vessel's signature. A remotely positioned reference sensor is used to remove variations in the Earth's background field that occur during the ship's calibration.

Undegaussed signatures and degaussing system loop effects can be measured much more rapidly and accurately with the ship moored inside a fixed facility in comparison to a degaussing range. By networking the ship's onboard digitally controlled degaussing system with the calibration facility's data acquisition and control computer, the induced and perm signatures, along with all loop effects, can be measured automatically and placed in a standard grid within a few minutes and before significant changes in tide (sensor depth) can occur. Conversely, ranging a ship requires 10–20 min for each undegaussed signature component and for each loop effect. Variations in the ship sensor geometry for each crossing of the range, and changes in tide that occur over the extended period needed to collect all the loop effects and undegaussed measurements, reduce the accuracy of the calibration data.

Ranges are more efficient for fine tuning a degaussing system's settings that are needed to compensate for changes in a vessel's perm. Degaussing ranges have been placed at entrances to naval harbors and ports to measure the magnetic signatures of fleet vessels as they routinely enter and leave. This allows periodic adjustments to be made in a degaussing controller without requiring dedicated time from the ship or interfering with its schedule. Degaussing ranges are especially useful in forward areas, located far from any fixed magnetic facility.

Once the loop effects and undegaussed signatures have been measured and extrapolated to a standard grid, degaussing loop ampere-turns are then computed to minimize the vessel's off-board fields. The first step in the calibration process is to place the undegaussed measurements, H, which have been corrected to a standard grid, into a column matrix such as:

$$H = \begin{bmatrix} h_1 \\ h_2 \\ h_3 \\ \vdots \\ h_n \end{bmatrix}, \qquad (3.1)$$

where h_n represents the nth discreet measurement of the ship's undegaussed field. The loop effects, C, also corrected to a standard grid and normalized to 1 ampere-turn, are placed into a rectangular matrix:

$$C = \begin{bmatrix} c_{1,1} & c_{1,2} & c_{1,3} & \cdots & c_{1,m} \\ c_{2,1} & c_{2,2} & c_{2,3} & \cdots & c_{2,m} \\ c_{3,1} & c_{3,2} & c_{3,3} & \cdots & c_{3,m} \\ \vdots & \vdots & \vdots & \ddots & \vdots \\ c_{n,1} & c_{n,2} & c_{n,3} & \cdots & c_{n,m} \end{bmatrix}, \qquad (3.2)$$

where m is the number of degaussing loops, and the nth discreet measurement of the loop effect must correspond to the same location and vector component as the nth term in the undegaussed field matrix, H.

Least-squares signature minimization is typically used in degaussing coil calibration computations since it is a deterministic and linear process in which capacity constraints can be placed on the ampere-turns of the loops. The basic equation to be minimized can now be expressed as:

$$(H + CI)^{\mathrm{T}} (H + CI) = \min, \qquad (3.3)$$

where I is a column matrix of m terms representing the ampere-turns for each loop that minimizes Equation (3.3). Solving Equation (3.3) in the least-squares sense gives:

$$I = \left[C^{\mathrm{T}} C \right]^{-1} C^{\mathrm{T}} H. \qquad (3.4)$$

As was discussed under the topic of inverse modeling in Holmes [6], Equation (3.4) is inherently unstable due to the physics of this specific problem. Although unstable source strength solutions may be acceptable when extrapolating signatures, wildly varying ampere-turns between adjacent degaussing loops can result in a significantly overspecified system that would unnecessarily increase the weight, volume, electric power demand, air conditioning load, and cost of the ship. In addition, measurement and extrapolation errors, combined with an unstable ampere-turn solution, can result in a less than optimal degaussed signature.

The least-squares degaussing loop calibration computation can be stabilized using the same techniques used for inverse mathematical modeling as explained by Holmes [6]. For degaussing system calibrations, the minimum energy stabilization criterion, as mathematically derived by Twomey [7], is most appropriate. This criterion forces the sum of the squares of the degaussing loops' ampere-turns to a minimum while simultaneously reducing the vessel's signature. The minimum energy constraint is applied to the degaussing calibration computation by modifying Equation (3.4) to give:

$$I = \left[C^{\mathrm{T}} C + \alpha i \right]^{-1} C^{\mathrm{T}} H, \tag{3.5}$$

where i is the identity matrix, and α is a weighting factor. In practice, the α term (called a damping factor) is adjusted empirically to minimize the ampere-turn demands for the degaussing loops, while also reducing the ship's signature to the specified field level.

The degaussing loop calibration computation given by Equation (3.5) is repeated for each of the four undegaussed signature components. That is, H is loaded with the undegaussed ILM measurements, and then in turn the IAM, IVM, and total perm (PLM + PAM + PVM) signatures. In practice, modern computerized degaussing systems control the current flowing to each loop, while holding the number of active conductors in each loop constant. For this reason, the loop current needed to degauss each induced signature are normalized to amperes per nT of applied inducing field and are stored in the onboard control computer's database. The current required to cancel the total perm signature is simply a constant offset applied to each loop and does not have to be scaled before saving it in the controller's database.

Although a naval vessel's magnetic field may be calibrated to a low level while the ship is at the degaussing range or inside a fixed facility, changes in its induced and permanent magnetization while at sea will degrade its well-degaussed signature. As discussed by Holmes [2], the three induced components of magnetization will change quickly with the ship's location (latitude and longitude), its heading, and its roll and pitch angles, whereas the permanent magnetization drifts more slowly due to the application of mechanical stress on the ferromagnetic hull, internal structure, and machinery items. Therefore, the onboard degaussing controller that regulates the current flowing to each degaussing loop must account for these changes and update them on a continual, real-time, basis to maintain a low signature.

A block diagram of an onboard degaussing controller is shown in Figure 3.8. The control loop starts in the upper left side of the diagram, where the system acquires the roll, pitch, and heading angles, and the present latitude and longitude from the ship's navigation system. This data serves as input to a computer model of the Earth's main magnetic field, called GEOMAG [8], whose output is the Earth's inducing field in the ship's frame of reference. Some vessels built with a nonmagnetic superstructure can use a triaxial magnetic field sensor mounted on its mast to measure

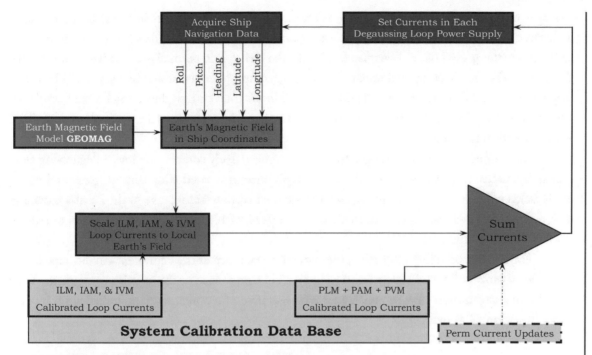

FIGURE 3.8: Flow diagram for controlling degaussing coil currents to maintain the compensation of a ship's ferromagnetic signature components.

the local earth's field directly in *ship coordinates* (ship's reference frame), which, in waters with high geologic magnetic anomalies, is more accurate than the global computer model.

After the local earth's magnetic field is computed or measured in ship coordinates, the normalized degaussing loop currents, retrieved from the onboard data bank at system startup, are scaled to the local field. The scaled currents for the three induced components are then summed, along with the perm currents stored onboard at calibration. The controller transmits a requested current setting to each individual degaussing loop's power supply and then restarts the process by acquiring a new set of data from the ship's navigation system. The cycle time for one update of the degaussing current is 10–100 ms, depending on the roll rate of the vessel.

Because the permanent magnetization of a naval vessel does change over time, due to mechanical stress and voyage effects, the degaussing controller must be updated periodically with new perm current settings. Typically, changes to the perm currents are made based on the analysis of the vessel's signatures measured with a degaussing range. The easiest approach to recalibrating a degaussing system is to assume that any variation in the measured signature, when compared to the vessel's last ranging, is caused by changes in its perm. (This approach does assume that the degaussing system's original induced settings are correct and that there are no errors in translating previously measured signatures to the present degaussing range environment.) Updates to the perm current settings are computed using Equation (3.5) with the changes in perm signatures inserted into the H matrix. The newly computed updates to the perm currents can be stored in a separate file in the degaussing controller's database and added to each loop's total current during each update cycle, as shown in Figure 3.8, or the perm current updates can be added directly to the original perm settings and restored in the database.

Roll-induced eddy current signatures can also be actively compensated with degaussing systems. As discussed in the previous chapter, a ship's uncompensated eddy current-generated magnetic fields will have both an in-phase and a quadrature component relative to the Earth's inducing field. Therefore, degaussing coils must also be energized with phase quadrature currents to reduce this signature component.

For this introductory discussion, the inductive reactance of the equivalent circuit impedance of the eddy current flow will be assumed to be small in comparison to the resistive component. This simplifying assumption may be considered representative of a nonconducting hulled ship. In this case, Equation (2.4) reduces to:

$$h_e \propto \frac{-j\omega AB_e \theta_{max} e^{j\omega t}}{R}.$$

(3.6)

It is clear from Equation (3.6) that, in this example, the ship's uncompensated eddy signature will have only a pure quadrature component. Therefore, the degaussing coil currents must also have a quadrature component.

Calibration of a ship's degaussing system to compensate roll-induced eddy current signatures takes place inside an Electromagnetic Roll Facility such as that shown in Figure 3.7b. The fixed facility's inducing coils are energized with alternating current (AC) at the natural roll frequency of the vessel under calibration. The magnitude of the current is selected so as to generate a peak inducing field comparable to the maximum roll angle expected for the ship class. The current flowing in the facility's inducing loops is monitored and used both as a phase reference for the calibration and to remove the AC inducing field's influence on the underwater array of magnetometers. Therefore, only the distortion in the AC inducing field caused by the presence of the ship is measured, separated into in-phase and quadrature signature components, and recorded by the facility.

Once the uncompensated quadrature component of the eddy current fields are measured and separated from the in-phase, the degaussing system's quadrature coil currents that will minimize this signature can then be computed. Placing the uncompensated quadrature eddy current signature into the H matrix of (3.5), the degaussing coil currents that will minimize this signature component can then be computed. Once again, the loop current needed to degauss the vertical and athwartship components of the eddy current signatures are normalized to amperes per nT of applied inducing field and are stored in the control computer's onboard database. However, in this case, the quadrature compensating current in each degaussing loop will have to be scaled not only by the ship's roll, pitch, heading, latitude, and longitude but also according to its roll frequency.

A ship's onboard degaussing controller must have the ability to set currents in each loop's power supply to cancel both the in-phase and quadrature signature components. By differentiating the temporal variation of the modeled or measured Earth's field, referenced in the ship's coordinates, a quadrature component of the inducing field can be produced that is also proportional to the ship's roll frequency (Figure 3.9). Scaling the normalized eddy current compensating currents stored in the degaussing controller's database during system calibration and adding them to the induced and perm control currents, both the roll-induced and ferromagnetic ship signature components can be minimized simultaneously.

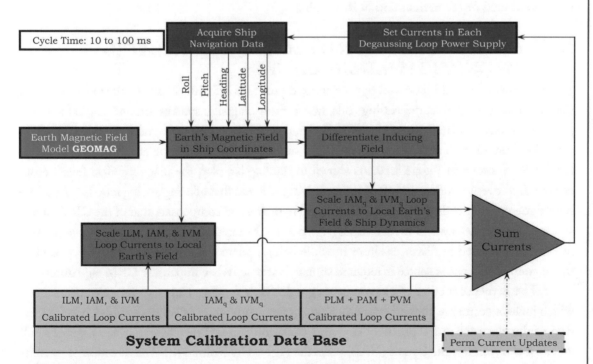

FIGURE 3.9: Flow diagram for controlling degaussing coil currents to maintain the compensation of a ship's ferromagnetic and roll-induced eddy current signature components.

Although, in principle, a naval vessel's global degaussing coil system can actively compensate the magnetic signatures produced by stray field sources, it is difficult, in practice, to maintain a low stray field signature. Omitting the possibility of a shipboard electric power system having a ground loop through the seawater, a major problem on its own, all stray field sources are composed of loops of current formed by their closed circuits. As a result, all stray field sources form magnetic dipoles and can be actively compensated with degaussing loops, which are also magnetic dipoles. However, the current amplitude and its physical path through the circuits of a ship's power system can change quickly. When this happens, the degaussing loop's current settings loaded into the controller at calibration are no longer correct for compensating the stray fields. The stray field signature component could then be undercompensated or overcompensated by the degaussing system.

The optimum method for reducing stray field signatures is through the proper up-front design of the electric power system's components or subsystems for minimal magnetic field. This approach was discussed in the previous chapter under passive signature reduction techniques. If necessary, dedicated small degaussing loops may be installed around a single major stray field source, such as a high-power motor or generator, but they must be controlled through monitoring of the item's local field or the current flow at its terminals.

3.3 ACTIVE REDUCTION OF CORROSION-RELATED MAGNETIC FIELD SIGNATURES

Since the magnetic field produced by an electric dipole falls off with distance at a slower rate than that of a magnetic dipole, degaussing coils, which form magnetic sources, cannot be used to compensate corrosion-related magnetic (CRM) signatures. The primary magnetic field produced by corrosion-related currents circulates around the surface ship or submarine according to the right-hand rule as drawn in Figure 3.10. As shown in the log–log plot, the magnetic field falloff of an electric dipole representing the CRM source is $1/R^2$, whereas that of a degaussing loop is $1/R^3$. If the source strength of the degaussing loop is adjusted to match and compensate that of the CRM source at a specific distance away from the vessel (e.g., point p_i in Figure 3.10), the resultant signature will be overcompensated at distances closer in while being undercompensated further away. It is clear that a controlled electric source is required to match and actively compensate CRM signatures.

The active reduction of electric or magnetic fields produced by a vessel's electric sources, which include corrosion and cathodic protection system currents, is called *deamping*. A brief discussion of the electrochemical process that forms corrosion currents that flow between a vessel's hull and its nickel–aluminum–bronze (NAB) propeller is given by Holmes [2]. The direction of conventional current inside a freely corroding steel hull and in the surrounding seawater is shown in Figure 3.11. Since the hull is anodic and can be thought of as a positive source that injects electric current

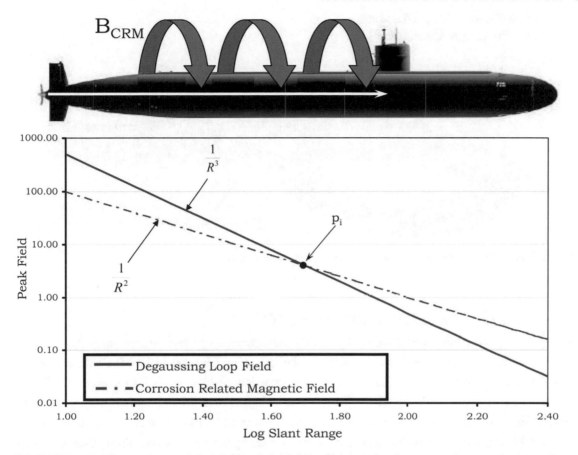

FIGURE 3.10: Comparison of the idealized field fall-off rates of a degaussing loop with that of the uncompensated corrosion-related magnetic source.

into the seawater, a logical approach to its compensation would be to artificially generate a negative source on the hull to cancel it. However, in doing so, the hull would corrode at an extremely accelerated rate in the neighborhood of the artificial cathode. Another approach is required.

A deamping system may only use positive sources of electric current on a vessel's steel hull to actively compensate its CRM signature, while simultaneously protecting it from corrosion. The anodes of an ICCP system meet the positive current requirements, and have been used extensively to cathodically protect large steel hulled commercial and naval vessels for many years. A boundary element modeling technique for optimizing the number and location of a ship's ICCP anodes to minimize its off-board electric field signature was presented by Diaz et al. [9]. In principle, a similar deamping technique could be developed for minimizing a vessel's CRM signature.

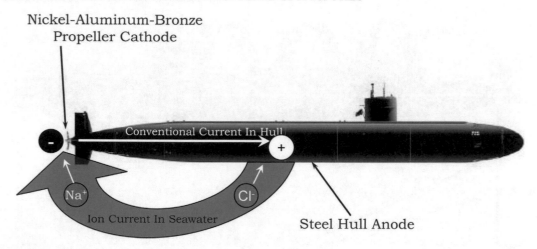

FIGURE 3.11: Current paths of a freely corroding naval vessel.

3.4   CLOSED-LOOP DEGAUSSING

A degaussing controller that assumes the vessel's permanent magnetization remains constant after calibration is called an *open-loop degaussing* (OLDG) system. Changes in PLM, PVM, and PAM signatures are caused by the mechanical stresses experienced by a ship's ferromagnetic hull and internal structure as it sails within the Earth's magnetic field. An OLDG system has no ability to detect these changes and recompensate the perm currents while the vessel is at sea. However, the perm field components can be kept within loose bounds by periodically measuring a vessel's magnetic signature at a degaussing range and, if needed, updating the perm currents to recalibrate the system.

Fixed and transportable degaussing ranges are used to keep the magnitude of a ship's magnetic signature within specified limits. The maximum field amplitude allowed for a naval vessel when measured at a standard depth is called the *check range limit*. If a ship's signature exceeds this limit, which is usually caused by changes in the perm components, its degaussing system must be recalibrated until the field amplitude is below a specified value called the *calibration limit*. In practice, the frequency of ranging a ship depends on naval policy and the availability of a degaussing range in the vicinity of its operating area.

As improvements are made to degaussing systems and the signature limits are lowered, periodic recalibration at a degaussing range is not sufficient to maintain the field below the check range limit. A degaussing control system is needed to continually monitor changes in a ship's permanent magnetization while it is underway and to automatically adjust the perm current settings to maintain a very low magnetic signature. Degaussing control of this type is called *closed-loop degaussing* (CLDG).

An OLDG system can be converted into a CLDG scheme through the addition of two subsystems. The first subsystem comprises a large array of magnetic field sensors (magnetometers) positioned throughout the vessel, which are used to measure changes in its permanent magnetization while at sea. The second subsystem is the data acquisition and transmission system that provides these measurements to the degaussing controller (Figure 3.12). The degaussing controller reads in the onboard magnetic field measurements, computes changes to the perm currents needed to reoptimize the signature, and then updates the perm current settings in the OLDG database.

The flow of the CLDG process is shown in Figure 3.12 (lower left box). First, the onboard magnetic field measurements are read into the control computer and are processed to remove noise and any known interference that may be present in the collected data. The onboard magnetic field associated with changes in the vessel's permanent magnetization is then extrapolated to equivalent changes in the off-board signature at the standard depth equivalent to a degaussing range. This step

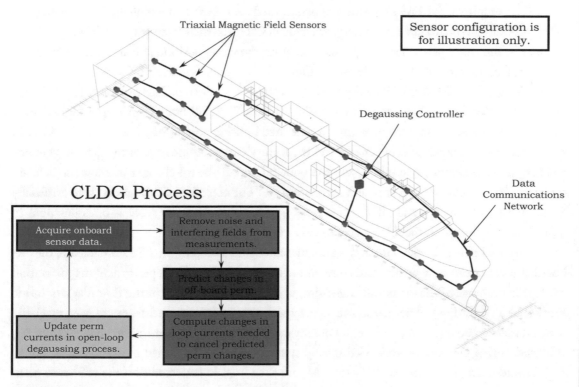

FIGURE 3.12: CLDG concept.

is necessary since the degaussing coils have been designed to reduce the ship's magnetic field at the standard depth and not on or inside its hull. The predicted changes in off-board perm signatures are then placed in Equation (3.5) to compute changes in the perm current settings needed to reduce them. The perm currents in the OLDG system are updated and the CLDG process repeats. The cycle time for updating perm currents by the CLDG process is much slower than the OLDG loop that controls the induced and eddy current components.

It is clear that the critical component of the CLDG process is the prediction of the off-board perm fields from onboard measurements. This is a difficult step to accomplish with the accuracy required by a CLDG system. The onboard sensors are located in the extreme near field of the vessel's flux pattern and are influenced by high-order source terms whose fields ironically may contribute little to the off-board signature at the standard depth. These high-order sources and fields are the primary reason that large numbers of onboard sensors are required to prevent aliasing of the spatially sampled fields. Equivalent source, empirical, and statistical models have been considered for predicting off-board signatures from onboard measurements. An empirical signature prediction technique will be discussed here.

The empirical CLDG signature prediction method is based on simultaneous measurement of the on- and off-board changes in magnetic field distribution associated with artificially produced changes in the ship's permanent magnetization. With the vessel positioned inside a fixed magnetic silencing facility, such as the one shown in Figure 3.7b, a direct current (DC) bias field is established along one of the axes while a deperming sequence is conducted. This process forces a change in magnetization in the direction of the bias field (Figure 3.13, left box). In principle, higher-order changes in permanent magnetization can be generated by using nonuniform bias fields. The CLDG system's string of magnetic field sensors and the facility's array of underwater fluxgate magnetometers are used to simultaneously measure the onboard and off-board changes in magnetic field associated with the change in permanent magnetization. The objective of this process is to artificially generate a set of onboard and off-board magnetic state vectors that can serve as basis functions to reproduce any change in the ship's magnetization while it is at sea.

After all perm change state vectors are measured and stored in the CLDG database, the onboard controller can now predict and compensate changes in the vessel's permanent magnetization while it is underway. The prediction algorithm is given mathematically in the flow diagram inside Figure 3.13 (right box). After noise and interference signals are removed from onboard magnetic measurements, the data are placed in a column matrix, $\vec{F}_{on}$. The onboard perm state vectors, $\vec{S}_{on}$, collected during calibration are loaded during system startup and are used in a least-squares fit to $\vec{F}_{on}$. The resulting computed scale factors, $\vec{M}$, are then used to scale and sum the off-board perm state vectors, $\vec{S}_{off}$, also collected during system calibration. The predicted change in off-board perm signatures, $\vec{F}_{off}$, is placed in the least-squares degaussing current computation of Equation (3.5).

a.) CLDG Calibration

b.) CLDG Operation

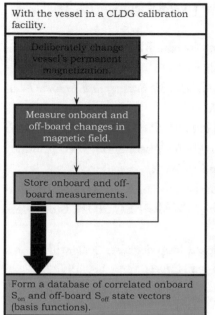

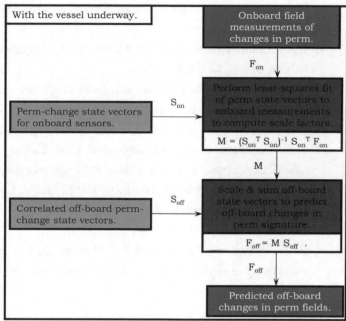

FIGURE 3.13: Flow diagram of the CLDG system's calibration process and its underway operation.

The perm current settings are updated in the OLDG degaussing controller, and the CLDG process then repeats.

Work is ongoing to improve the CLDG process [10]. The objective of this research is to reduce the number of onboard sensors and, if possible, eliminate the need for dedicated ship time during system calibration. Reducing system costs and ship impact while improving performance will increase the attractiveness of CLDG systems.

REFERENCES

[1] M. F. Schoeffel, A short history of degaussing, Bureau of Ordinance, NAVORD OD 8498, Washington, D.C., Feb. 1952.

[2] J. J. Holmes, *Exploitation of a Ship's Magnetic Field Signatures*, 1st edn. Morgan & Claypool Publishers, San Rafael, CA, 2006. doi:10.2200/S00034ED1V01Y200605CEM009

[3] M. Norgen and S. He, Exact and explicit solution to a class of degaussing problem, *IEEE Trans. Mag.*, 36(1), Jan. 2000.

[4] K. R. Davey, Degaussing with BEM and MFS, *IEEE Trans. Mag.*, 30(5), Sep. 1994. doi:10.1109/20.312681

[5] F. LeDorze, J. P. Bongiraud, J. L. Coulomb, P. Labie, and X. Brunotte, Modeling of degaussing coils effects in ships by the method of reduced scalar potential jump, *IEEE Trans. Mag.*, 34(5), Sep. 1998.

[6] J. J. Holmes, *Modeling a Ship's Ferromagnetic Signatures*, 1st edn. Morgan & Claypool Publishers, San Rafael, CA, 2007.

[7] S. Twomey, *Introduction to the Mathematics of Inversion in Remote Sensing and Indirect Measurement.* Dover Publications, Inc., Mineola, NY, 1977.

[8] S. McLean (2007. Feb.). NGDC GEOMAG Version 6.0 (10/2005). Geomagnetic Models and Software. National Geophysical Data Center, NOAA. Washington, DC. [Online]. Available: http://www.ngdc.noaa.gov/seg/geomag/models.shtml.

[9] E. S. Diaz, R. A. Adey, J. Baynham, and Y. H. Pei, Optimization of ICCP systems to minimize electric signatures, *Proc. Marine Electromagn. Conf.* (MARELEC 2001), Stockholm, Sweden, 2001.

[10] R. A. Wingo, M. Lackey, and J. J. Holmes, Test of closed-loop degaussing algorithm on a minesweeper engine, *Proc. Am. Assoc. Naval Engineers Conf.*, Crystal City, VA, 1992.

• • • •

CHAPTER 4

Summary

Reducing a naval vessel's susceptibility to actuating an influence mine or a submarine's detectability to surveillance systems by modifying their magnetic field signatures is called magnetic silencing. In deeper waters, it is possible to lower a naval vessel's signature to a level where an influence bottom mine cannot detect it, eliminating the need to hunt or sweep mines in these areas during the initial time-critical and resource-demanding stages of the conflict. In shallower waters, reducing magnetic signatures has the effect of increasing sweeping efficiency caused by forcing the minefield planner to increase his mines' sensitivity in order to maintain a desired threat level. Magnetic signature reduction decreases the effective density of the minefield, lessening the time and platform resources needed to clear the minefield and lowering the risk to follow-on ship traffic.

Naval areas of interest have shifted toward littoral scenarios and away from deep water operations. The acoustically challenging shallow-water ocean environments have increased the importance of detecting submarines by their electromagnetic field signatures. Small low-power magnetic field sensors can now be deployed in underwater submarine barrier arrays and in manned or unmanned aircraft. Swarms of magnetic anomaly detection (MAD)-equipped unmanned air vehicle (UAV) controlled in a cooperative behavior search pattern could monitor large shallow-water areas of the ocean and detect acoustically quiet submarines. In addition to protecting it from magnetic influence mines, decreasing a submarine's magnetic signature will also reduce its susceptibility to detection by these surveillance systems.

There are four primary shipborne sources of magnetic field in the ultralow frequency band that ranges from approximately zero to 3 Hz: ferromagnetism, roll-induced eddy currents, corrosion-related currents, and currents flowing in electric power systems. The most important shipboard source of magnetic field is the induced and permanent magnetization caused by the interaction of the Earth's natural magnetic field with the ferromagnetic steel used in the construction of a naval vessel's hull, internal structure, machinery, and equipment items. The second most important shipboard source of magnetic fields are eddy currents induced in any electrically conducting materials including nonmagnetic metals, such as aluminum or stainless steels, which are generated primarily when the ship rolls within the Earth's magnetic field. The third largest, and least known, of the major sources of magnetic field, is electrochemically generated corrosion currents or cathodic protection system currents that flow between anodic and cathodic areas along the vessel. The last of the major shipboard sources of magnetic field are produced by large DC and alternating current (AC) flowing in the electric circuits found in high-power electric generators, motors, switchgear, breakers, and the distribution cables that interconnect them. Almost every aspect of ship and ship system design can affect underwater electromagnetic field signatures.

Before attempting to actively cancel any ship signature, elimination of its source should be pursued to the maximum degree possible within technical and affordability constraints. A naval vessel's ferromagnetic source strength can be significantly reduced by using less magnetic steel in its construction. This passive signature reduction technique is implemented by decreasing the size of the vessel or using materials with a low magnetic permeability such as austentic stainless steel, aluminum, titanium, carbon composites, or fiberglass. The latter two materials, having significantly lower electrical conductivity than the others, will also reduce the roll-induced eddy current field. Reduction of the stray field sources caused by current flowing in electric power systems is best accomplished through proper up-front design of individual components or power subsystems with the purpose of reducing their magnetic field. The corrosion-related current sources can be decreased by using electrochemically similar materials in the vessel's construction, through the application of coatings and paints to those dissimilar metals exposed to the seawater, and by decreasing the current output of the cathodic protection system as much as possible. These passive signature reduction techniques have a much lower impact on a naval vessel's performance and operation than active signature compensation systems.

Any magnetic field component that requires further reduction after the application of passive techniques can be compensated by artificially generating a flux distribution whose pattern is the negative of the residual field. Degaussing was the first active signature cancellation system and was developed during World War II to compensate a naval vessel's ferromagnetic signature component. A degaussing system is composed of a triaxial set of cable loops, which when energized with the proper current will cancel the field produced by the ship's induced and permanent magnetization. In principle, a degaussing system can also compensate the magnetic fields produced by roll-induced eddy currents and stray fields generated by electric power systems. However, a degaussing system, which is a magnetic source, cannot effectively cancel the fields originating from electric sources produced by corrosion or cathodic protection system currents. Corrosion-related signatures are actively reduced with a deamping system, which is composed of impressed current cathodic protection system anodes that have been positioned on a ship's hull to minimize the vessel's signature.

The major drawback of all active signature reduction systems is the requirement to monitor changes in the naval vessel's source strengths and to reoptimize the compensation system's settings. Adjusting the current setting in an active reduction system to minimize ship's signature is called calibration. Calibration is carried out by sailing the ship across a permanent or portable degaussing range, or with the vessel moored above a fixed array of magnetic field sensors, while system settings are changed to minimize the measured off-board fields.

Degaussing systems are typically calibrated with the least squares minimization algorithm. The individually measured loop effects form a set of basis functions that are used in the calculation to determine changes to loop currents that will minimize the signature's root-mean-squared magnitude. Because of the mathematically ill-conditioned nature of the active signature reduction calibration process, additional constraints must be incorporated into the solution to avoid oscillatory current settings of large amplitude. If the degaussing system calibration computation is not stabilized, an overdesigned system could result with significant adverse impacts on the ship. Typically, a minimum energy constraint is incorporated into the degaussing system calibration process.

Once a vessel has been calibrated and leaves the facility, it must monitor and maintain its well-degaussed state throughout normal operations. Since a ship's induced magnetization changes with its orientation within the Earth's local magnetic field, a degaussing system controller must monitor its location and orientation. The hull's roll, pitch, heading, latitude, and longitude serve as input to a mathematical model of the Earth's magnetic field to determine the triaxial inducing fields in the ship's coordinate system. Ships with nonmagnetic superstructures can use triaxial magnetic field sensors mounted on their mast to measure the inducing field directly. The degaussing loop currents established during calibration are then scaled to the local Earth's field and set in the system at an update rate of 10–100 Hz. In principle, the roll-induced eddy currents can also be compensated with a ship's degaussing system if the inducing field's rate of change is measured or computed. Stray field signatures are not generally compensated with the vessel's global degaussing system due to the very large number of possible circuit configurations and loads, which are nearly impossible to track and compensate in real time.

Changes in a naval vessel's permanent magnetization are the most difficult of the important magnetic field sources to monitor and keep well compensated with a degaussing system. Frequent ranging and recalibration of a degaussing system will prevent the perm signature from drifting too far astray from specifications. However, operational demands of the vessel generally prevent it from ranging and recalibrating its degaussing system at a desirable periodicity.

A closed-loop degaussing system can monitor changes in a naval vessel's permanent magnetization and automatically recalibrate the degaussing system in real time while underway. A closed-loop system is composed a large number of shipboard magnetic field sensors and associated data acquisition hardware networked to the degaussing controller. The controller uses the onboard measurements and a signature prediction algorithm to estimate changes in the off-board perm signature. (The perm signature extrapolation is necessary since the degaussing coils have been designed to optimally cancel the field at a specified depth beneath the vessel rather than onboard.) The extrapolated delta in the perm signature is then placed in the standard least-squares algorithm to compute the changes in perm current settings that will reminimize the off-board fields. After the perm current settings are updated, the process is repeated.

Future improvements to magnetic field signature reduction technologies will focus on reducing system costs and impact on the vessel's performance and operation. In the near term, the application of high-temperature superconducting degaussing cable has the potential of reducing system weight by a factor of 10 and cost by more than a factor of 2 [1]. In the far term, passive signature reduction techniques not only have the greatest potential for significantly lowering field levels and increasing the war-fighting capabilities of naval vessels but also may be the best approach for cutting total ship ownership costs. The ultimate goal of underwater electromagnetic signature research and development (R&D) is to develop technologies that can make a combatant undetectable by influence mines or surveillance systems.

REFERENCE

[1] High temperature superconductor degaussing system, Degaussing Data Sheet, American Superconductor, 2 Technology Dr., Westborough, MA 01581, 2006.

. . . .

Author Biography

John J. Holmes received his B.S. (1973), M.S. (1974), and Ph.D. (1977) degrees in electrical engineering from the West Virginia University. He joined the Naval Surface Warfare Center (1977) and is currently the Senior Scientist for the Underwater Electromagnetic Signatures and Technology Division, where he is responsible for the development of underwater electromagnetic field signature reduction systems for surface ships and submarines. Dr. Holmes has written three books and 27 peer-reviewed papers and holds 12 patents. He received the 2006 Naval Sea System Command Scientist of the Year Award, the David Packard Excellence in Acquisition Award (1999), and the Meritorious Civilian Service Award (1986). Dr. Holmes is a senior member of the Institute of Electrical and Electronic Engineers.

Printed in the United States
by Baker & Taylor Publisher Services